Cell Growth and Apoptosis

The Practical Approach Series

★ indicates new and forthcoming titles

Affinity Chromatography
Anaerobic Microbiology
Animal Cell Culture (2nd Edition)
Animal Virus Pathogenesis
Antibodies I and II
★ Basic Cell Culture
Behavioural Neuroscience
Biochemical Toxicology
★ Bioenergetics
Biological Data Analysis
Biological Membranes
Biomechanics — Materials
Biomechanics — Structures and Systems
Biosensors
★ Carbohydrate Analysis (2nd Edition)
Cell–Cell Interactions
★ The Cell Cycle
★ Cell Growth and Apoptosis
Cellular Calcium
Cellular Interactions in Development
Cellular Neurobiology
Clinical Immunology
Crystallization of Nucleic Acids and Proteins
★ Cytokines (2nd Edition)
The Cytoskeleton
Diagnostic Molecular Pathology I and II
Directed Mutagenesis
★ DNA Cloning 1: Core Techniques (2nd Edition)
★ DNA Cloning 2: Expression Systems (2nd Edition)
★ DNA Cloning 3: Complex Genomes (2nd Edition)
★ DNA Cloning 4: Mammalian Systems (2nd Edition)
Electron Microscopy in Biology
Electron Microscopy in Molecular Biology
Electrophysiology

Enzyme Assays
Essential Developmental Biology
Essential Molecular Biology I and II
Experimental Neuroanatomy
★ Extracellular Matrix
★ Flow Cytometry (2nd Edition)
Gas Chromatography
Gel Electrophoresis of Nucleic Acids (2nd Edition)
Gel Electrophoresis of Proteins (2nd Edition)
★ Gene Probes 1 and 2
Gene Targeting
Gene Transcription
Glycobiology
Growth Factors
Haemopoiesis
Histocompatibility Testing
★ HIV Volumes 1 and 2
Human Cytogenetics I and II (2nd Edition)
Human Genetic Disease Analysis
Immunocytochemistry
In Situ Hybridization
Iodinated Density Gradient Media
★ Ion Channels
Lipid Analysis
Lipid Modification of Proteins
Lipoprotein Analysis
Liposomes
Mammalian Cell Biotechnology
Medical Bacteriology
Medical Mycology
★ Medical Parasitology
★ Medical Virology
Microcomputers in Biology
Molecular Genetic Analysis of Populations
★ Molecular Genetics of Yeast
Molecular Imaging in Neuroscience
Molecular Neurobiology
Molecular Plant Pathology I and II
Molecular Virology
Monitoring Neuronal Activity
Mutagenicity Testing
★ Neural Cell Culture
Neural Transplantation
Neurochemistry
Neuronal Cell Lines
NMR of Biological Macromolecules
★ Non-isotopic Methods in Molecular Biology
Nucleic Acid Hybridization
Nucleic Acid and Protein Sequence Analysis
Oligonucleotides and Analogues
Oligonucleotide Synthesis
PCR 1
★ PCR 2
★ Peptide Antigens
Photosynthesis: Energy Transduction
★ Plant Cell Biology
★ Plant Cell Culture (2nd Edition)
Plant Molecular Biology

★ Plasmids (2nd Edition)
Pollination Ecology
Postimplantation Mammalian Embryos
Preparative Centrifugation
Prostaglandins and Related Substances
★ Protein Blotting
Protein Engineering
Protein Function
Protein Phosphorylation
Protein Purification Applications
Protein Purification Methods
Protein Sequencing
Protein Structure
Protein Targeting
Proteolytic Enzymes
★ Pulsed Field Gel Electrophoresis
Radioisotopes in Biology
Receptor Biochemistry
Receptor–Ligand Interactions
RNA Processing I and II
Signal Transduction
Solid Phase Peptide Synthesis
Transcription Factors
Transcription and Translation
Tumour Immunobiology
Virology
Yeast

Cell Growth and Apoptosis

A Practical Approach

Edited by

GEORGE P. STUDZINSKI

Department of Laboratory Medicine and Pathology,
UMDNJ-New Jersey Medical School,
Newark, NJ

Oxford New York Tokyo

Oxford University Press, Walton Street, Oxford OX2 6DP
Oxford New York
Athens Auckland Bangkok Bombay
Calcutta Cape Town Dar es Salaam Delhi
Florence Hong Kong Istanbul Karachi
Kuala Lumpur Madras Madrid Melbourne
Mexico City Nairobi Paris Singapore
Taipei Tokyo Toronto
and associated companies in
Berlin Ibadan

Oxford is a trade mark of Oxford University Press

Published in the United States
by Oxford University Press Inc., New York

A catalogue record for this book is available from the British Library

Library of Congress Cataloging in Publication Data
Cell growth and apoptosis: a practical approach/edited by George P. Studzinski. – 1st ed.
(The Practical approach series; no. 159)
Includes bibliographical references and index.
1. Cells–Growth–Research–Methodology. 2. Apoptosis–Research –Methodology. I. Studzinski, George P. II. Series.
QH604.7.C445 1995 574.87'61–dc20 95–3378

ISBN 0 19 963569 2 (Hbk)
ISBN 0 19 963568 4 (Pbk)

Typeset by Footnote Graphics, Warminster, Wilts
Printed in Great Britain by Information Press Ltd, Eynsham, Oxon.

Preface

The contributors to this volume have combined to present in simple language the detailed methodologies required to characterize diverse aspects of cell growth and division. The focus of this book is on the analysis of growth of normal and neoplastic human cells in culture. This anthropocentric bias is the result of the impossibility of covering important aspects of all life forms in a book of this type, and is consistent with Alexander Pope's reflection that 'the proper study of mankind is man'. Thus no apology is offered for this bias, and each chapter in this book is designed to facilitate both basic and clinical studies of human cells. The currents that I anticipate will carry us into the 21st century include automated methods for measurement of all facets of cell growth, such as flow cytometry, and mathematical analysis of experimental results. Also, the recent recognition that growth has not only a positive but also a negative dimension is reflected in the current popularity of the term 'apoptosis'. Thus DNA damage, analysis of cell death, and even cell-mediated cytotoxicity have been included in the broad connotation of 'growth'.

The preparation of this volume was a real team effort. I thank all the authors, the publishing staff, Sumant Ramachandra and Claudine Marshall for their invaluable help with the manuscript, and Christa, who gave up her dining room table for three months while it was covered with manuscripts. I do hope the reader will profit from all this.

New Jersey
October 1994

G. S.

Contents

List of contributors xix

Abbreviations xxiii

1. Measuring parameters of growth 1
Renato Baserga

1. Introduction 1

2. Growth parameters 2
Indirect measurements of cell number 3
DNA amount 3
RNA amount 4
Protein amount 4

3. Mitoses 4

4. DNA synthesis 7
Autoradiography 7

5. Determination of parameters of growth by autoradiography with [^{3}H]thymidine 12
Growth fraction 12
Quiescent cells and stationary cell populations 13
Phases of the cell cycle 14
Short method for growth fraction and cell cycle phases 15
Flow of cells through the cell cycle 16

6. Quiescence of cells 17

References 18

2. Cell cycle analysis by flow cytometry 21
Harry A. Crissman

1. Introduction 21

2. Cell preparation and fixation 22

3. DNA-specific fluorochromes commonly used for staining 23
DNA reactive Hoechst dyes, DAPI and DIPI 24
Mithramycin, chromomycin, and olivomycin 24
Propidium iodide (PI), ethidium bromide (EB), 7-amino-actinomycin D, and hydroethidine (HE) 24
Acridine Orange 25
TOTO-1, YOYO-1, TO-PRO-1, and YO-PRO-1 25

4. Cell staining 25
Viable cell staining with Hoechst 33342 25
Staining of fixed cells and unfixed nuclei 26

5. Instrumentation setup and analysis 27
Standards 27
Application 28

6. Simultaneous DNA and protein assay 28
Cell staining 29
Instrument setup and analysis 29
Applications 30

7. Cellular bromodeoxyuridine and DNA contents 30
Differential fluorescence analysis of BrdUrd content 32
BrdUrd pulse labelling 32
Cell staining with Hoechst 33342 (HO) and mithramycin (MI) 33
Instrument setup for differential fluorescence analysis 34
Analysis of stained cells 34

8. Correlated, three-colour, DNA–RNA–protein determinations 36
Cell staining 37
Three-laser FCM setup for analysis of DNA–RNA–protein 37
Analysis of three-colour stained cells 38
Controls for optimal colour resolution 39

9. Correlated analysis of cyclin B1 and cell cycle 40
Cell staining 40
Analysis by flow cytometry 41

Acknowledgements 42

References 42

3. Analysis of flow cytometric DNA histograms 45
Peter S. Rabinovitch

1. Introduction 45

2. Histogram analysis 45

3. Cell cycle modelling 48
The DNA content cell cycle 48
Model components 50
Debris modelling 51
Aggregation modelling 53
Complexity of the model 54

4. Reliability of histogram analysis results 56
CV and peak shape 56
Quantitation of aggregates and debris 56
Proportion of aneuploid cells 57

Statistical measures derived from the fitting model 57
Sensitivity of S and G_2/M phase measurements to variations in the fitting model 57

References 58

4. Determination of the cell proliferative fraction in human tumours 59

Rosella Silvestrini, Maria Grazia Daidone, and Aurora Costa

1. Introduction 59

2. Methods for procurement and preparation of tumour specimens 60
Procurement of tumour tissue 60
Preparation of specimens from solid tumours 60
Treatment of cell aspirates 61

3. Proliferation indices 62
Incorporation of nucleic acid precursors 62
Other measurements of cell proliferation 69
Methodological controls 73
Relationship between the different proliferation indices 73

4. Predictive value of cell proliferative fraction measurements 74
Prognosis 75
Response to treatment 76

Acknowledgements 76

References 77

5. Chromosome analysis in cell culture 79

Anwar N. Mohamed and Sandra R. Wolman

1. Introduction 79

2. Blood and bone marrow cultures 80
Blood 80
Bone marrow optimal culture conditions 81
Methotrexate synchronization 81

3. Solid tissue cultures 85
Culture conditions of solid tumours 85

4. Staining and banding methods 88
Trypsin–Giemsa staining (G-banding) 89
Quinacrine staining (Q-banding) 89
Reverse (R-banding) 90
Constitutive heterochromatin banding (C-banding) 91

5. Nomenclature and interpretation 92
Numerical and structural aberrations 92
Tumour cell populations 93

6. *In situ* hybridization and interphase cytogenetics 93

References 96

6. Assessment of DNA damage in mammalian cells by DNA filter elution methodology 97

Richard Bertrand and Yves Pommier

1. Introduction 97

2. DNA lesions measured by filter elution assays 97

3. Basic principles of the DNA filter elution methodology 98

4. Overview of the DNA filter elution procedures 100
Equipment 100
DNA labelling and preparation of experimental cell cultures 101
Calibration and internal standard cells 101
Cell lysis, DNA elution procedures, and treatment of filters 103

5. Specific conditions for each of the DNA filter elution assays 104
Filters and solutions 104
Single- and double-strand breaks 106
Protein-associated strand breaks and protein-free breaks 106
DNA–protein crosslinks 106
Interstrand DNA crosslinks 107
Alkali-labile sites 107
Apoptosis-associated DNA fragmentation 107
Preparation of isolated nuclei for filter elution assays 108
Reconstituted cell-free system with filter elution assays 108

6. Computations 109
SSB (single-strand breaks) 109
DSB (double-strand breaks) 111
DPC (DNA–protein crosslinks) 112
ISC (interstrand crosslinks) 113
Apoptosis-associated DNA fragmentation 113

7. Conclusions and perspectives 114

Acknowledgements 115

References 116

7. Morphological and biochemical criteria of apoptosis 119

Sumant Ramachandra and George P. Studzinski

1. Introduction 119

2. Distinction of apoptosis from other forms of cell death 122

3. Morphological changes in apoptosis 122
Identification by light microscopy of cells undergoing apoptosis 125
Identification by electron microscopy of cells undergoing apoptosis 128

4. Histochemical detection of apoptosis 129

5. Biochemical detection of specific DNA damage for demonstration of apoptosis 131
Determination of the increased mitochondrial to nuclear DNA ratio for detection and quantitation of apoptosis 136

6. Assessment of nuclear proteolytic activity in cells undergoing apoptosis 139

7. Concluding remarks 140

Acknowledgements 140

References 140

8. Analysis of cell death by flow cytometry 143

Zbigniew Darzynkiewicz, Xun Li, Jianping Gong, Shinsuke Hara, and Frank Traganos

1. Introduction 143

2. Modes of cell death 143
Apoptosis 144
Necrosis 145
Atypical apoptosis: delayed reproductive, or mitotic, cell death 145

3. Features discriminating dead cells 146
Structural and functional integrity of the plasma membrane 146
Cell organelles 147
Specific features of apoptotic cells 147

4. Light scattering properties of dying cells 148

5. Removal of dead cells by incubation with DNase I and trypsin 148

6. Exclusion of PI combined with hydrolysis of fluorescein diacetate 150

7. Uptake of Rh123 combined with exclusion of PI 151

8. Exclusion of PI followed by counterstaining with HO342 153

9. Uptake of HO342 combined with exclusion of PI 155

10. Extraction of the degraded DNA from apoptotic cells 156

11. Denaturability of DNA *in situ* 158

12. Labelling of DNA strand breaks in apoptotic cells 162
DNA strand labelling with biotinylated dUTP 162
DNA strand break labelling with digoxygenin-conjugated dUTP 163

13. Which method to choose? 164

Acknowledgements 166

References 166

9. Assays of cell growth and cytotoxicity 169

Philip Skehan

1. Introduction 169

2. Growth and cytotoxicity assays 169

3. Experimental design 170

4. Cell cultures 171
Seeding density 171
Dissociation and recovery 171
Drug solubilization 171
Assay duration 171
Control wells on every plate 172

5. Dye-binding assays 172
Optimizing and validating dye-binding assays 172
Protein and biomass stains 173

6. Metabolic impairment assays 179
Neutral Red viability assay 179
MTT cell viability assay 180

7. Membrane integrity assays 181
Fluorescein diacetate 181

8. Electrical conductivity assays 183

9. Survivorship assays 184
Adherent versus non-adherent cells 185
Long-term recovery (LTR) assay 185
Colony-forming efficiency on tissue culture plastic 187
Colony-forming efficiency in soft agar 188

References 190

10. Cell synchronization as a basis for investigating control of proliferation in mammalian cells 193

Gary S. Stein, Janet L. Stein, Jane B. Lian, T. J. Last, Thomas Owen, and Laura McCabe

1. Introduction 193

2. Synchronization of continuously dividing cells by thymidine 193
General considerations 193
Rationale for steps in thymidine synchronization protocol 197
Modification of thymidine synchronization for specialized applications 198
Precautions that should be exercised during thymidine synchronization 198

3. Synchronization of continuously dividing cells by mitotic selective detachment 200
General considerations 200
Modification of mitotic selective detachment protocols 200

4. Induction of synchronous proliferative activity 202

References 203

11. Growth and activation of human leukaemic cells *in vitro* and their growth in the SCID mouse model 205

Sophie Visonneau, Alessandra Cesano, and Daniela Santoli

1. Introduction 205

2. Procedures for growing human leukaemic cells in suspension culture 206

3. Clonal growth of leukaemic cells in semi-solid medium 208
Choice of the semi-solid medium 209
Preparation of the agar stock 209
Preparation of the methylcellulose stock 211
Scoring of colonies in semi-solid medium 212
Staining of colony-forming cells 212

4. Use of the SCID mouse model for growing human leukaemias 212
The SCID mouse 212
Immunosuppressive treatments of SCID mice 214
Transfer of leukaemic cells i.p. 215
Intravenous (i.v.) injection of leukaemic cells 215

Success of human leukaemic cell engraftment 215
Appearance of symptoms 216
Autopsy and histopathology 216
Recovery of human leukaemic cells from mouse tissues 217

5. Human lymphocyte activation in culture 218
Cell-mediated cytotoxicity 218
Expression of intracytoplasmic azurophilic granules and lytic proteins 219
E/T conjugate formation 219
Direct lysis of target cells 220
Role of Ca^{2+} in the killing process 222
Exocytosis of cytotoxic granules 223
Proliferative responses 225
Lymphokine production 225

Acknowledgements 225

References 226

12. Senescence and immortalization of human cells 229

Karen Hubbard and Harvey L. Ozer

1. Introduction 229

2. Cellular senescence 230

3. SV40-transformation and crisis 235

4. SV40-immortalized cell lines 238

5. Conditional SV40 transformants 241

6. Other approaches to immortalization of human cells 243

7. Summary 244

References 244

Appendix. Changes identified in senescing mammalian cells 245

13. Analysis of the effects of purified growth factors on normal human fibroblasts 251

Paul D. Phillips and Vincent J. Cristofalo

1. Introduction 251

2. Studies with growth factors 253
Materials 253
Trypsinizing and harvesting the cells 255
A classification of growth factors for WI-38 cells 257

References 262

A1. *Addresses of suppliers* 265

Index 267

Contributors

RENATO BASERGA
Department of Microbiology and Immunology, Jefferson Cancer Institute, 233 South 10th Street, Philadelphia, PA 19107, USA.

RICHARD BERTRAND
Centre de recherche L. C. Simard, Institut du Cancer de Montreal, Hopital Notre-Dame, 1560 Sherbrook St East, Montreal, Quebec, Canada H2L 4M1.

ALESSANDRA CESANO
The Wistar Institute, 3601 Spruce Street, Philadelphia, PA 19104, USA.

AURORA COSTA
Oncologia Sperimentale C, Istituto Nazionale per lo Studio e la Cura dei Tumori, Via Venezian 1,20133 Milan, Italy.

HARRY A. CRISSMAN
Cell Growth, Damage and Repair Group, Los Alamos National Laboratory, Los Alamos, NM 87545, USA.

VINCENT J. CRISTOFALO
Center for Gerontological Research, Medical College of Pennsylvania, 2900 Queen Lane, Philadelphia, PA 19129, USA.

MARIA GRAZIA DAIDONE
Oncologia Sperimentale C, Istituto Nazionale per lo Studio e la Cura dei Tumori, Via Venezian 1,20133 Milan, Italy.

ZBIGNIEW DARZYNKIEWICZ
The Cancer Research Institute, New York Medical College, Valhalla, NY 10595, USA.

JIANPING GONG
The Cancer Research Institute, New York Medical College, Valhalla, NY 10595, USA.

SHINSUKE HARA
The Cancer Research Institute, New York Medical College, Valhalla, NY 10595, USA.

KAREN HUBBARD
Department of Microbiology and Molecular Genetics, UMDNJ–New Jersey Medical School, 185 S. Orange Avenue, Newark, NJ 07103, USA.

T. J. LAST
Department of Cell Biology, University of Massachusetts Medical Center, 55 Lake Avenue North, Worcester, MA 01655, USA.

XUN LI
The Cancer Research Institute, New York Medical College, Valhalla, NY 10595, USA.

JANE B. LIAN
Department of Cell Biology, University of Massachusetts Medical Center, 55 Lake Avenue North, Worcester, MA 01655, USA.

LAURA McCABE
Department of Cell Biology, University of Massachusetts Medical Center, 55 Lake Avenue North, Worcester, MA 01655, USA.

ANWAR N. MOHAMED
Pathology Department/Cytogenetics, Wayne State University/Harper Hospital, 3990 John R., Detroit, MI 48201, USA.

THOMAS OWEN
Department of Cell Biology, University of Massachusetts Medical Center, 55 Lake Avenue North, Worcester, MA 01655, USA.

HARVEY L. OZER
Department of Microbiology and Molecular Genetics, UMDNJ–New Jersey Medical School, 185 S. Orange Avenue, Newark, NJ 07103, USA.

PAUL D. PHILLIPS
Biology Department, West Chester University, West Chester, PA 19383, USA.

YVES POMMIER
Laboratory of Molecular Pharmacology, Bldg. 37, Rm 5C25, NCI, NIH, Bethesda, MD 20892, USA.

PETER S. RABINOVITCH
Department of Pathology SM-30, University of Washington, Seattle WA 98185, USA.

SUMANT RAMACHANDRA
Department Laboratory Medicine and Pathology, UMDNJ–New Jersey Medical School, 185 S. Orange Ave., Newark, NJ 07103, USA.

DANIELA SANTOLI
The Wistar Institute, 3601 Spruce Street, Philadelphia, PA 19104, USA.

ROSELLA SILVESTRINI
Oncologia Sperimentale C, Istituto Nazionale per lo Studio e la Cura dei Tumori, Via Venezian 1,20133 Milan, Italy.

PHILIP SKEHAN
Andes Pharmaceuticals, Inc. P.O. Box 598, Los Altos, CA 94023–0598, USA.

GARY S. STEIN
Department of Cell Biology, University of Massachusetts Medical Center, 55 Lake Avenue North, Worcester, MA 01655, USA.

JANET L. STEIN
Department of Cell Biology, University of Massachusetts Medical Center, 55 Lake Avenue North, Worcester, MA 01655, USA.

GEORGE P. STUDZINSKI
Department Laboratory Medicine and Pathology, UMDNJ–New Jersey Medical School, 185 S. Orange Ave., Newark, NJ 07103, USA.

FRANK TRAGANOS
The Cancer Research Institute, New York Medical College, Valhalla, NY 10595, USA.

SOPHIE VISONNEAU
The Wistar Institute, 3601 Spruce Street, Philadelphia, PA 19104, USA.

SANDRA R. WOLMAN
Pathology Department/Cytogenetics, Wayne State University/Harper Hospital, 3990 John R., Detroit, MI 48201, USA.

Abbreviations

ActD	actinomycin D
ADCC	antibody-dependent cell-mediated cytotoxicity
AdR	adriamycin
ALL	acute lymphoblastic leukaemia
AML	acute myelogenous leukaemia
ALS	alkali-labile sites
ANLL	acute non-lymphoblastic leukaemia
AO	Acridine Orange
ATCC	American Tissue Culture Collection
ARA-C	1-β-arabinocytosine
AZRA	Azure A
BAD	background aggregates and debris
BCGF	B-cell growth factor
BLT	*N*-benzyloxycarbonyl-L-lysine thiobenzylester
BM	bone marrow
BMCM	bone marrow culture medium
BPB	Bromophenol Blue
BSA	bovine serum albumin
BrdUrd	bromodeoxyuridine
CA-P	calcium phosphate
CD	cluster designation
CAM	camptothecin
CEE	chick embryo extract
CFE	colony-forming efficiency
CFU	colony-forming unit
CHO	Chinese hamster ovary
ChA3	chromomycin
CK	creatine kinase
CM	conditioned medium
CPDL	cumulative population doubling level
CSA	colony-stimulating activity
CSF	colony-stimulating factor
CTL	cytotoxic T lymphocytes
CTR	chromotrope 2R
CV	coefficient of variation
CVS	chorionic villi
DAPI	4′,6-diamidino-2-phenylindole
DEX	dexamethasone
DI	DNA index

DIPI	4,6 bis-(2 imidozolynyl-4H, 5H)-2-phenylindole
DMEM	Dulbecco's modified Eagle's medium
DMSO	dimethylsulfoxide
DOC	deoxycholate
DPC	DNA–protein crosslinks
DSB	double strand breaks
EB	ethidium bromide
EBSS	Earle's balanced salt solution
EBV	Epstein–Barr virus
EBVNA–2	EBV nuclear antigen 2
EGF	epidermal growth factor
EGTA	ethylene bis(oxyethylenenitrilo)-tetraacetic acid
E/T	effector cell: target cell ratio
ETOH	ethanol
FB	frank breaks
FBS	fetal bovine serum
FCM	flow cytometric
FCS	fetal calf serum
FDA	fluorescein diacetate
FGF	fibroblast growth factor
FITC	fluorescein isothiocyanate
FISH	fluorescence *in situ* hybridization
F–P	Ficoll–Paque
GAGs	glycosaminoglycans
GCT	giant cell tumour
GM	granulocyte–macrophage
GM–CSF	granulocyte–macrophage colony-stimulating factor
HBSS	Hank's balanced salt solution
HDL	high density lipoprotein
HE	hydroethidine
HF	human fibroblasts
HS	horse serum
HO	Hoechst
HPV	human papilloma virus
HSF	human skin fibroblast
HTLV	human T-cell leukaemia virus
IGF	insulin-like growth factor
IL	interleukin
IMDM	Iscove's modified Dulbecco's medium
INS	insulin
ISC	interstrand DNA crosslinks
LAK	lymphokine-activated killer
LCM	lymphocyte-conditioned medium
LPD	lymphoproliferative disorders

LPS	lipopolysaccharides
LTR	long-term recovery, or long terminal repeat
mAb	monoclonal antibody
β-ME	β-mercaptoethanol
MEM	minimum essential medium
MHC	major histocompatibility complex
MI	mithramycin
MLR	mixed lymphocyte reaction
MLV	murine leukaemia virus
MMCT	microcell-mediated cell fusion
MMTV	mouse mammary tumour virus
MSA	multiplication stimulating activity
MTT	3-(4,5-dimethylthiazole-2-yl)-2,5-diphenyl tetrazolium bromide
MW	molecular weight
MTX	methotrexate
NK	natural killer
OD	optical density
ORG	Orange G
PAI	plasminogen activator inhibitor
PB	peripheral blood
PBL	peripheral blood lymphocytes
PBS	phosphate-buffered saline
PCA	perchloric acid
PCNA	proliferating cell nuclear antigen
PD	population doublings
PDGF	platelet-derived growth factor
PFP	pore-forming protein
PHA	phytohaemagglutinin
PI	propidium iodide
PPP	platelet-poor plasma
PPLO	pleuropneumonia-like organism
PWM	pokeweed mitogen
PY	pyronin
QE	quantum efficiencies
RBC	red blood cells
RSV	Rous sarcoma virus
SCF	stem cell factor
SCID	severe combined immunodeficient
SE	serine esterases
SSB	single-strand breaks
SSC	saline–sodium citrate
SFME	serum-free mouse embryo
SOD	Mn-superoxide dismutase

SPF	S-phase fraction
SRB	sulforhodamine B
STI	soybean trypsin inhibitor
STSP	staurosporine
SV40	simian virus-40
T-ALL	T lymphoblastic leukaemia
TBO	Toluidine Blue O
TCA	trichloroacetic acid
TdT	terminal deoxynucleotidyl transferase
THR	thrombin
TNN	thionin
TNE	buffer containing Tris, NaCl, and EDTA
TOTO-1	thiazole orange dimer
TPA	tetradecanoylphorbol-13-acetate
TRS	transferrin
YOYO-1	Oxazole yellow dimer

1

Measuring parameters of growth

RENATO BASERGA

1. Introduction

This chapter deals with the various parameters of growth and how they can be measured.

A tissue can grow by: (i) increasing the number of cells; (ii) increasing the size of the cells; or (iii) increasing the amount of intercellular substance. Since the intercellular substance of a tissue is usually a secreted product of the cell, for example collagen, it can be considered as an extracellular extension of the cytoplasm. We can therefore consider an increase in intercellular substance as a variation of an increase in cell size and thereby reduce tissue growth to two mechanisms, growth in size and growth in the number of cells. This is true regardless of whether we are dealing with normal or abnormal growth, with the intact animal or with cells in culture. However, although both mechanisms may be operative, an increase in cell number is (with very few exceptions) by far the most important component in either normal or abnormal growth (*Table 1*).

There are static and dynamic ways of measuring growth and cell division. For instance, counting the number of cells in a Petri dish tells us how much that cell population has grown. It does not tell us whether or not the cells are still proliferating. Other methodologies (autoradiography with [^{3}H]thymidine, flow cytometry, immunohistochemistry, etc.) are necessary if we wish to examine cell proliferation and its perturbations in more detail.

Table 1. Postnatal growth of rat liver

Age (days)	Liver wt. (g)	Cell number ($\times 10^{-6}$)	DNA (pg/cell)	Protein (pg/cell)
10	0.30	168	5.9	29.3
21	0.98	445	5.1	45.6
41	5.7	1060	11.1	103.0
80	8.1	1270	11.4	154

Adapted from ref. 1.

The following sections give a few simple techniques for measuring cell growth, and stress their interpretation and their limitations. The choice of the techniques that will be described below was dictated in part by their practicality and in part by the fact that other techniques (for instance, flow cytometry) have been made available in Chapter 2 of this volume.

2. Growth parameters

From the foregoing, it is clear that there are several ways of measuring parameters of growth. The question is: which parameter of growth does one wish to measure? Take, for instance, a typical experiment in which one wishes to determine whether a certain growth factor is mitogenic for a population of cells in culture. Mitogenic means that it induces mitosis: that is, that cells divide and increase in number. It is wrong to measure mitogenicity by a labelling index (with [^{3}H]thymidine or with bromodeoxyuridine) or, even worse, by incorporation of radioactive thymidine into acid-soluble material. Incorporation of [^{3}H]thymidine measures DNA synthesis, not cell division. The two processes often go together, but they can also be separated (for a review, see ref. 1). If we wish to determine the effect of growth factors (or of any environmental change) on cell proliferation, that is their ability to stimulate or inhibit cell division, the best method is very simple: to count the number of cells before and after treatment, possibly at 24-hour intervals.

Protocol 1. Counting the number of cells in cultures

1. Prepare the dishes in which the cells have been grown (this example is for 100-mm dishes).
2. Prepare a trypsin solution, either 0.25 or 0.1% [a] in Hanks' balanced salt solution (containing no Ca^{2+} or Mg^{2+}).
3. Pour Hanks' solution (no Ca^{2+} or Mg^{2+}) into 50-ml tubes.
4. Remove the medium from the dishes and set aside.
5. Wash with 10 ml of Hanks'; remove.
6. Add 10 ml of trypsin solution and leave for 30 sec to 1.5 min. [a]
7. Remove the trypsin solution and let stand at room temperature for 2–3 min. [a]
8. Add the growth medium (which includes 5% calf serum), 10 ml [b]. At this point the cells will detach from the surface. In these days of very expensive serum one can use the conditioned medium obtained from step (4) to inhibit trypsin instead of fresh growth medium. Be careful, however, that in certain experiments (apoptosis is very fashionable these days), there may be a lot of dead cells floating in the growth medium; in this case, fresh growth medium is required.

9. Mix well using a sterile pipette, drawing the cell suspension up and down the pipette 5–10 times.
10. Count in a haemocytometer, by depositing a few drops of cell suspension under the coverslip[c]. Use the four corners to count cells, divide by 4 and multiply by 10^4 to obtain cells/ml. For instance, if you count 140 cells in the four corners: 140/4 = 35; 35×10^4 per ml, and since the cells had been collected in a 10 ml suspension = 3.5×10^5 cells per dish.

[a] Trypsin strength varies from one batch to another (regardless of what the manufacturer says) and, in addition, sensitivity to trypsin is different in different cell lines. Therefore, it is impossible to give a single optimal trypsin concentration. One has to go by trial and error, and the same comments apply to steps 6 and 7. The goal is to obtain a suspension of single cells with as little cellular debris as possible.
[b] The amounts can be appropriately scaled down if one uses smaller Petri dishes. The amounts of growth medium we use to *grow* cells (not to trypsinize them) are 20, 8, and 3 ml for 100-, 60- and 35-mm respectively. In short-term experiments (24–48 h), the amounts of growth medium can be reduced to half the indicated volumes, resulting in considerable savings.
[c] Counting the number of cells in a solid tissue is somewhat more complicated. The difficulty here is to obtain a satisfactory suspension of single cells. Enzymatic dissociation of the tissue is helpful, but tricky, and has to be adapted to each different type of tissue. Perhaps, for solid tissues, determination of DNA amount (see Section 2.2) is preferable.

2.1 Indirect measurements of cell number

The number of cells per dish is logically the best parameter to use to indicate whether or not a population of cells is growing. There are, however, alternatives. For instance, one can measure the amount of DNA per dish or per gram of tissue, or the amount of RNA, or the amount of protein, or count the number of mitoses.

2.2 DNA amount

The amount of DNA per dish can be determined, for instance, at 24-h intervals after plating. It is a very simple procedure; indeed, for someone trained in biochemistry, it is easier than counting the number of cells. Since the amount of DNA per cell is *usually* constant (in mammalian cells, 6×10^{-12} g/diploid cell in G_1), the amount of DNA per dish is an indirect measure of the number of cells. The G_1 amount of DNA in somatic cells is generally referred to as the $2n$ amount. Cells in S-phase or in G_2 have increased amounts of DNA with respect to G_1 cells, but, if a population is truly growing, the increase in total amount of DNA per dish will go way beyond the error caused by the individual variations due to the distribution of cells throughout the cell cycle. Determination of DNA amount is probably the method of choice for solid tissues. Using the amount of DNA per cell given above, one can calculate (from the amount of DNA per μg of tissue) the number of cells per μg of tissue. A rough (very rough) estimate is that

1 g of tissue contains 5×10^8 cells, but this estimate will vary greatly with the type of tissue.

A possible source of error is polyploidy; that is, an increase in the amount of DNA per cell from $2n$ to $4n$ or even $8n$, in which case one could have an increase in the amount of DNA per dish without a concomitant increase in the number of cells. This is not a frequent occurrence, but it can happen, for instance, in certain pathological conditions, in response to certain drugs, or when a large proportion of cells is blocked in G_2 (reviewed in ref. 1). A real problem with tissues, regardless of the method used to determine the number of cells, is that tissues are a mixture of various types of cells. For instance, in human tumours, the tumour cells constitute 50–60% of all cells, the others being connective tissue cells, blood vessels, inflammatory cells, etc. Clearly, one would like to know the number of *tumour cells*, but this is not easy, even by flow cytometry, unless one is lucky enough to have tumour cells with a markedly different ploidy from the the normal cells. One should therefore be extremely careful in evaluating cell growth parameters in solid tissues.

There are several methods for determining the amount of DNA in a culture dish or in a tissue. I prefer the classic method of Burton (2).

2.3 RNA amount

For a given cell type, the amount of RNA ought to be constant. As with DNA, G_2 cells will have roughly twice the amount of RNA as G_1 cells. As a measure of cell number, however, RNA amount is not accurate because, in several conditions, cells can grow in size and increase their RNA amount without cell division (3). RNA amount is a good indicator of cell size, rather than cell number. Most of the cellular RNA that is measured by bulk chemical methods or individual cell histochemical methods is ribosomal RNA (rRNA ~85% of total cellular RNA). Since rRNA forms a part of the ribosome, on which protein synthesis is carried out, it seems logical that RNA amounts ought to be a reasonable indicator of cell size, a hypothesis that has been empirically confirmed. Classic methods for the determination of RNA amounts can be found elsewhere. If one wishes to determine the amount of RNA in individual cells, flow cytometry is by far the best method, since both RNA and DNA amounts can be determined simultaneously on the same batch of cells. Flow cytometry is discussed in Chapter 2.

2.4 Protein amount

The same comments apply here as to RNA amount. The amount of cellular protein is a reasonable indicator of cell size, but a poor indicator of cell number (4).

3. Mitoses

Surely, if one is looking at the mitogenic effect of a substance, the number of mitoses ought to be the best indicator of such an effect. However, mitoses are fleeting; in most cells they last only 45 min and, unless one looks at precisely the right moment, one may miss them. Furthermore, the duration of mitosis can increase in certain cells, especially in transformed cells (5). Everything else being equal, if the duration of mitosis in cell line A is twice that of cell line B, the number of mitoses in A will also be twice that of B, although the two cell lines may grow at the same rates. In tissues, the number of mitoses per 1000 cells (the mitotic index) is a reasonable measure of the *proliferating activity* of a cell population, but not of its growth. For instance, in the crypts of the lining epithelium of the small intestine, there are many mitoses. Fortunately for us, the small intestine in the adult individual does not grow, because, for every new cell produced in the crypts, one dies at the tips of the villi. So, in any given cell population, one must distinguish between cell division and increase in cell number. Even in growing tumours, cell death is a frequent occurrence (1), and the mitotic index of a tumour is only an inaccurate indication of its aggressiveness.

There are also technical problems. To begin with, if we wish to determine the mitotic index of cells in culture, a 22-mm^2 coverslip will have to be placed into the culture dish (see *Protocol 2* for the preparation of coverslips). Mitoses can then be counted directly on the coverslips after fixation and staining (see below). Staining and counting of mitoses directly on plastic surfaces is not advisable. However, suppose that a wave of mitoses occurs 25–27 h after stimulation of a quiescent cell population with growth factors. One may miss it, unless samples are taken practically every hour. More economic in terms of time and money is to add a drug that will arrest cells in mitosis. One can then count the percentage of cells that *accumulate* in mitosis over a certain period of time. The three main drugs for this purpose are colcemid, colchicine and nocodazole.

Colcemid and colchicine are very similar but the latter is more toxic. For mitotic arrest, the optimal concentration of colcemid is 0.16 μg/ml for human cells or 0.04–0.08 μg/ml for rodent cells. I like to leave the drug in for 4 h, then fix the coverslips. By dividing the time period into 4-h blocks (for instance, 16–20 h; 20–24 h; 24–28 h) after serum-stimulation, one should be able to get a pretty good idea of the mitotic activity of a cell population. If cells are left in colcemid (and especially colchicine) for more than 4 h, cell damage occurs with loss of mitotic figures. As an amusing aside on the fallacity of the mitotic index to measure cell growth, one should remember that 60 years ago, colchicine was considered to be mitogenic . . . because, when colchicine was given to animals, the number of mitoses markedly increased. The lesson we can draw from it is still valid today, since abnormal mitoses (frequent in tumours) can often be blocked, and thus give a false impression on the proliferative activity of the particular tumour.

Nocodazole (6) offers the advantage that it can be used for longer periods of time, 16–24 h. We use it at concentrations of 0.04–0.2 μg/ml, and mitoses, clearly identifiable, continue to accumulate. Depending on the cell line, and up to 12–16 h, nocodazole arrest is reversible (so is colcemid-arrest but only up to 4 h).

Protocol 2. To clean coverslips for tissue cultures

1. Pour chromic sulphuric acid (enough to cover the coverslips) into a large Petri dish.
2. One by one, place each coverslip in the acid. Leave to soak for 30–45 min.
3. Remove the acid and place the coverslips in a beaker. Let water run over them for 1–2 days.
4. After 1–2 days rinsing, rinse again with deionized water.
5. Then take three large Petri dishes. Fill the first with methanol and the other two with 80% ethanol. Place all the coverslips in methanol; then individually rinse each coverslip—first in one dish of ethanol, then in the other. Lay them out to dry on paper towels or wipe with gauze pads. Then autoclave. All handling after the chromic sulphuric acid is done with tweezers.
6. Coverslips are placed in tissue culture and, at the desired times, they are removed, washed three times in buffer and then fixed in methanol at −20°C for 15 min. The coverslips are then mounted (cells up!) on a regular glass slide for convenient handling, using ordinary nail polish.

Please note: this may seem a cumbersome way to clean coverslips. However, if the coverslips are not perfectly clean, fixatives may detach cells and cause a lot of problems. Especially for autoradiography and immunohistochemistry, scrupulously clean coverslips are a must.

Protocol 3. Cells arrested in mitosis by nocodazole

1. Coat the coverslips (four per 100-mm Petri dish) with poly-L-lysine (Sigma, 3000 mol. wt) at 1 mg/ml dissolved in Hanks' (calcium, magnesium free solution). Leave for 24 h.
2. Remove the polylysine solution and allow the coverslips to dry.
3. Plate 5×10^5 cells per 100-mm dish in normal growth medium, each dish containing two polylysine-coated coverslips.
4. After 18 h remove the medium and add fresh growth medium plus 0.1–0.2 μg/ml of nocodazole dissolved in dimethylsulphoxide (DMSO).
5. Leave for the desired period of time and then fix using the method given above and stain with Giemsa/Sorenson's buffer.

4. DNA synthesis

It is often desirable to measure DNA synthesis instead of cell proliferation. This is especially true if one wishes to study G_1 events leading to the replication of DNA or if one wishes to know the fraction of proliferating cells in a given cell population. Measurements of DNA synthesis are also much more impressive than counting cell number. For instance, a mitogenic stimulus may double the number of cells in a Petri dish in 24 h. In the same time, the fraction of cells labelled by [^{3}H]thymidine will go from 0.1 to 90%, virtually eliminating the need of statistical analysis.

The method of choice here is still high-resolution autoradiography with [^{3}H]thymidine. I will first outline the technique and then discuss its advantages and disadvantages. Many investigators, these days, prefer immunohistochemistry with antibodies to bromodeoxyuridine which, like thymidine, is incorporated into DNA. The two methods give comparable results. Bromodeoxyuridine has the advantage that it is not radioactive (but tritium is not a dangerous radioactive isotope); [^{3}H]thymidine is more quantitative, especially if one wishes to study the dilution of the label in prolonged experiments. For most situations, the cell cycle analysis detailed below is substantially the same, whether one used radioactive thymidine or antibodies to bromodeoxyuridine.

4.1 Autoradiography

The procedure is given in great detail in *Protocols 4–6* (modified from ref. 7); it is meant for beginners who have never done autoradiography. The precautions are actually excessive. In practice, after a few successful attempts, the operator will have gained enough confidence to allow himself/herself shortcuts and a certain amount of casualness. In the end, autoradiography will become a trivial technique.

Protocol 4. Autoradiography: preparing tissue sections, and dripping techniques

- Prepare coverslips for autoradiography as outlined in *Protocol 2*.
- Grow cells on coverslips and fix as described in *Protocol 3*.

A. *Preparing tissue sections for autoradiography*

1. Mount the section on precleaned slides rubbed just prior to use with fresh egg albumin (egg white). (Avoid commercial albumin since it contains phenol, a reducing agent which causes a high background.)
2. Cut tissue sections of 3–10 (usually 5) μm.

Protocol 4. *Continued*

3. Deparaffinize slides; set up 11 staining dishes:
   - (a–d) xylene—5 min each.
   - (e) 50% xylene and 50% absolute ethanol—5 min.
   - (f–j) 100% alcohol—3–5 min, then 95, 70, 50, 35%, always for 3–5 min.
   - (k) distilled water—until slides are ready to be dipped. Do not allow to stand in water longer than 0.5–1 h.

B. *Dipping technique*

Equipment

- Slides to be dipped carrying a coverslip or tissue section complete with numbers applied with Indian ink at least 1 h before dipping, plus two trial slides
- Slide boxes with bags of Drierite (Bakelite slide boxes, 25-slide capacity), Drierite, gauze sponges
- NTB2 (Nuclear Track Emulsion) (Eastman Kodak Co.)
- Scotch Brand Pressure Sensitive Tape (Minnesota Mining & Manufacturing Co.)
- X-ray film envelopes, or aluminium foil
- Red china marker
- L-shaped galvanized tray
- Timer
- Coplin jar half-filled with distilled water at 40°C
- Glass stirring rod
- Scissors
- Plastic bag
- Log (batch no., slide no., date dipped, date developed, date stained, type emulsion, etc.)
- Water bath at 40°C
- Darkroom—[use only Kodak Safe-Light 'Wratten' Series red lamp with 15-W bulb, with red filter, only when necessary]

Method

1. Melt the emulsion by placing it into constant temperature bath at 40°C for approximately 90 min.
2. Hydrate the tissue slides or mounted coverslips in distilled water not more than 0.5 h before dipping (see above).
3. If the emulsion is to be used undiluted, pour it into a beaker and keep it in a water bath at 40°C.
4. If the emulsion is to be used diluted: in complete darkness, except for a safe-light, fill the rest of the Coplin jar (see *Equipment*) with emulsion. (This 1:1 dilution may be discarded at the end of the experiment.)
5. DO NOT LET ANY LIGHT FALL ON EMULSION (turn safe-light toward wall).
6. Dip in two clean trial slides to test the consistency of emulsion.
7. Dip the experimental slides back to back, for about 2 sec, vertically with frosted ends up into emulsion. Separate the slides and place on an L-shaped tray, frosted ends forward and up. Be careful not to scrape the side of the emulsion container or touch the surface of slide; it may cause mechanical exposure.

8. Slides should air-dry in 20–30 min, but may take longer in a small damp room. Test trial slides to see if they are dry.
9. When the slides are completely dry, place 10 or less in a Bakelite box. Seal the closure edge with black tape. Wrap the box securely in light-tight film envelope or two layers of aluminium foil, then completely seal with black tape. Write the batch number (as given in the experimental protocol) with a china marker on the tape. Place the boxes in a plastic bag, wrap it around them and fasten it closed.
10. Place the batch in a refrigerator to allow for exposure.

4.1.1 Exposure

The exposure time of autoradiographs depends on the type and amount of isotope used. Mouse tissues treated with 10 μCi of [^{3}H]Tdr per mouse require about 10–12 days exposure time. Cell cultures: 0.02 μCi/ml of [^{3}H]thymidine for 24 h labelling require 3–4 days exposure. For short labelling pulses (30 min or so) use 0.5 μCi/ml and 3 days exposure. If you are in a hurry, simply increase the concentration of [^{3}H]thymidine, but remember, long exposure of cells to high concentrations can cause radiation damage. Do not forget that some animals (squirrels, woodchucks and others) and some cell lines do not have thymidine kinase activity (1). In these situations, one could use [^{3}H]deoxycytidine.

Protocol 5. Autoradiography: developing

Equipment

- Water bath at 18°C including two buckets of crushed ice and thermometer
- Darkroom, using only a 15-W bulb with red filter
- Six staining dishes placed in a water bath
- Staining trays, timer
- Distilled water
- D-19 Developer, make up fresh every week, store in a brown bottle at room temperature, dilution 595 μg in 3.8 litres, or 156 μg in 1 litre. Always filter before use (Eastman Kodak Co.)
- F-10 Fixer, make every 3–4 weeks and store in a brown bottle at room temperature (97 μg/ml) (Eastman Kodak Co., 1 lb package—3800 ml H_2O at room temperature), filter and dilute 1:1 with distilled water before use

Method

1. Darken room.
2. Fill the staining dishes with changes solutions (see below).
3. Place the slides in racks.
4. Slightly dirty the changes solutions by running an empty tray through them (it sounds magical but it works better).
5. Change solutions every 10 slides.

Protocol 5. *Continued*

6. Changes: (a) developer (*undiluted*)—*5 min*; (b) distilled water—brief rinse; (c) diluted fixer—8 min; (d) distilled water—5 min.
7. Dry (air-dry); store in dust-proof boxes until ready for staining.

Protocol 6. Autoradiography: staining

Here two techniques are given; one using haematoxylin–eosin and one using Giemsa, both of which are useful for cells in culture, smears or tissue sections.

A. *Giemsa staining*

1. Place well-rinsed autoradiographs in buffered distilled water (pH 6.8) for 2 h; leave to dry.
2. Stain with Giemsa, pH 4.8, for 1 h (Giemsa stock solution, 2.5 ml; methanol, 3 ml; distilled water, 100 ml; 0.1 M citric acid, 11 ml; 0.2 M disodium phosphate, 6.0 ml).
3. Rinse, air-dry and mount.

B. *Haematoxylin and eosin staining*

Equipment

NB: It is important that ALL reagents be filtered before use.

- *Mayer's haematoxylin:* 1 μg haematoxylin (mol. wt 356.34, Eastman Organic Chemicals) plus 0.2 μg $NaIO_3$ (mol. wt 197.901, Fisher Scientific Co.) plus 50 μg ammonium alum (aluminium ammonium sulphate, mol. wt 906.688, Fisher Scientific Co.), plus 1000 ml H_2O; will last about 2–3 months
- *Eosin Y:* stock solution (Eosin Y: water- and alcohol-soluble, Fisher Scientific Co.) 5.0 g, plus 1000 ml of H_2O. For use dilute 1 ml of stock solution with 99 ml of H_2O
- 1% sodium acetate: $NaC_2H_3O_2$ (Baker's Reagents, mol. wt 82.04), 10 g, plus 1000 ml of H_2O
- Water bath at 18°C (will need ice)
- Staining dishes and trays

Method

1. Prepare an 18°C water bath with 9 staining dishes. All reagents must be cooled to 18°C before procedure can begin.
2. Never use more than five slides per tray.
3. Remember to filter all reagents before use.
4. Do not use the same haematoxylin for more than two trays (10 slides). Change everything after one tray.
5. Every time a batch of autoradiographs is stained, the optimum time of staining is determined by use of a standard slide, which is stained first. The other slides can be stained only after this optimum time is determined.

6. Changes:
   (a–c) Soak autoradiographs in three changes of distilled water for 10 min each (total 30 min).
   (d) Stain with haematoxylin for 10–12 min. Check the intensity of stain under microscope. If not sufficient, stain and check again after another 1 min. Usually the staining time with haematoxylin will be found to vary between 2 and 3 min.
   (e) Rinse in water.
   (f) Blue in 1% sodium acetate for 5 min. Check under microscope. Repeat procedure if bluing is not satisfactory.
   (g) Rinse in water.
   (h) Stain with eosin for 1 min. Check under microscope. Repeat if necessary.
   (i) Rinse in water and air-dry.
7. Remove excess emulsion from back of slide with razor blade.
8. OPTIONAL: if autoradiographs are to be mounted, place slides in xylene, mount with Permount and coverglass.

4.1.2 Glossary

Trial slides are blanks used only to check when the emulsion has melted and when it is dry after dipping. A standard slide is a coverslip (or smear) of cells that have been labelled with [^{3}H]thymidine and *are known* to give good autoradiographs. For instance, once a year, we label a batch of HeLa cells with [^{3}H]thymidine for 24 h; a sample is taken and autoradiographed. If the autoradiography is of good quality, we make many slides carrying these HeLa cells and place one slide in all batches of autoradiographs to be done. This standard slide will serve as a monitor to check that the autoradiography procedure has been carried out correctly. We find that the use of standard slides decreases the occurrence of frustration. Again, after the operator has reached a sufficient level of competence, the standard slides can be omitted.

4.1.3 Pitfalls

Be careful of artefacts. Chemical or mechanical blackening of the emulsion can occur. The latter can be easily identified by its random distribution; the former can be more tricky. Formalin, for instance, if not carefully washed out, can give autoradiographic images that are quite convincing (but the cytoplasm will also be covered by grains). A high background is not bothersome if the radioactive label over the nuclei is very strong, but can become a problem if the label is weak. At any rate, good experimental practice suggests that one should always strive for a high quality end product. With tissue sections, airlocks can mimic an autoradiographic image.

5. Determination of parameters of growth by autoradiography with [^{3}H]thymidine

The proliferating activity of a cell population can be estimated in a qualitative way by the thymidine index; that is, by the percentage of cells labelled by pulse exposure to [^{3}H]thymidine. However, the thymidine index, like the mitotic index, gives only a rough estimate of the amount of cell proliferation in a tissue or in a cell population. For a more precise study, it is desirable to measure other parameters. The three cell cycle parameters that are based on the use of [^{3}H]thymidine and autoradiography are:

- the growth fraction
- cell cycle analysis
- the time in the cell cycle where a certain agent or manipulation arrests the cells

The way of determining these three parameters will be described separately for cells in culture, or in the living animal. Essentially the same analysis can be carried out with antibodies to bromodeoxyuridine.

5.1 Growth fraction

The fraction of cells in a cell population that are actively participating in the proliferative process can be determined *in vivo* (8) or *in vitro* [(9) tissue cultures] by continuous labelling with [^{3}H]thymidine.

5.1.1 *In vitro*

In exponentially growing cell populations the growth fraction is close to 100% and gives very little information. Thus, if we take almost any cell line that grows with a doubling time of 24 h or less during the exponential growth phase, and we label it with [^{3}H]thymidine for 24 h, the number of labelled cells would be close to 100%. The odd unlabelled cells can probably be considered as dying cells.

The fraction of labelled cells becomes important in experiments dealing with the effect of growth factors on the state of quiescence of a cell population. For instance, many cell lines stop proliferating when they reach confluence. It is desirable to test the extent of quiescence in a cell population in culture by exposing the cells to [^{3}H]thymidine. The interval usually taken, more for expediency than for any logical reason, is 24 h. Populations of cells that can be made quiescent should, at this point, have very low labelling indices; that is, not more than 1% of the cells ought to be labelled by a 24-h exposure to [^{3}H]thymidine. At this point if one wishes to determine the effect of growth factors on the proliferation of this particular cell population, all one has to do is to add the growth factors together with [^{3}H]thymidine and let the experiment run for 24 to 48 h. Cells entering S-phase will incorporate [^{3}H]thymidine and the thymidine index will reach a maximum point beyond which it will not increase. Usually, under these conditions, 80–90% of the cells can be stimulated to enter S-phase by appropriate stimuli. There is always a refractory fraction of cells that do not enter S-phase and this may be due to the fact that these cells have left the proliferative pool permanently or, alternatively, that the density of the population is such that they cannot be stimulated. In either case the experiment is very simple and it consists of comparing the fraction of cells labelled by [^{3}H]thymidine over the same period of time, 24 or 48 hours, in unstimulated cultures and in cultures stimulated with growth factors.

5.1.2 *In vivo*

Obtaining the growth fraction in living animals is more complicated. A simple injection of [^{3}H]thymidine is not sufficient with living animals because in the animal the injected [^{3}H]thymidine is promptly broken down in the liver, and an injection of [^{3}H]thymidine is practically equivalent to a 30–45 min pulse in a tissue culture. To obtain a reasonable growth fraction in mice or rats one can either use repeated injections of [^{3}H]thymidine, usually at intervals of 4 h or continuous infusion. Both methods are feasible when the cell cycle of the dividing fractions of cells is reasonably short and when the amount of radioactivity injected is not objectionable. The maximum number of cells labelled by continuous exposure represents the growth fraction. The growth fraction is very important in animal experiments. At variance with tissue cultures (exponentially growing cells), the growth fraction in tissues of the intact animal is far from approaching 100%. In fact, even fast-growing tumours may have growth fractions of only 10–20%. Without knowing the growth fraction, a kinetic analysis of tumour growth *in vivo* is meaningless.

5.2 Quiescent cells and stationary cell populations

It is important to distinguish quiescent populations of cells and stationary populations of cells. Stationary populations of cells are defined as populations of cells in which the number of cells does not increase. In ordinary circumstances this also means that the growth fraction would be very low and that very few cells would be labelled by [^{3}H]thymidine over a period of 24 or 48 h, but this is not always true. There are certain cell lines, especially transformed cell lines, that reach a stationary phase; that is, the number of cells increases no further, but if they are labelled for 24 h with [^{3}H]thymidine, one may find that up to 50% of the cells are labelled. This is often a characteristic of cells that are transformed with certain DNA oncogenic viruses but in general it is a characteristic of transformed cells. Puzzlingly enough, this mystery has never been solved. Some scientists believe that this is due to the fact that there is a continuous death of transformed cells in dense culture which are replaced by proliferating cells. Other scientists believe that the incorporation of [^{3}H]thymidine into stationary populations of transformed cells simply reflects an increase in the amount of DNA per cell without cell division. There is evidence that both mechanisms may actually be operative. Whatever the cause, the investigator ought to be cognizant of the fact that in certain instances there is a huge discrepancy between the number of cells that are labelled by [^{3}H]thymidine and the increase in the number of cells.

5.3 Phases of the cell cycle

To determine the duration of the various phases of the cell cycle, the following experiment is given as an illustration.

(i) Grow replicate cultures of BALB/c-3T3 cells on coverslips in Dulbecco's minimum essential medium supplemented with 10% fetal calf serum, glutamine and the usual antibiotics.

(ii) 48 h after plating, when the cells are still growing exponentially, expose all cultures to [^{3}H]thymidine at a final concentration of 0.2 μCi/ml.

(iii) After 30 min wash all cultures twice with balanced salt solution and then re-incubate in non-radioactive growth medium (see above).

(iv) At various intervals after removal of [^{3}H]thymidine, terminate groups of two or three cultures, remove the medium and wash and fix the coverslips for autoradiography, as described above.

When cultures thus treated are examined, no mitoses are found to be labelled immediately after removal of [^{3}H]thymidine, whereas about 35–40% of interphase cells are labelled. The information we have obtained is that DNA synthesis occurs only during interphase and not during mitosis. For several hours after removal of [^{3}H]thymidine, mitoses are still unlabelled, and since mitosis lasts for only about 45 min, there must be a period before mitosis during which cells do not synthesize DNA. This period is called the G_2 period, or post-synthetic gap. The length of the G_2 period is given by the interval between the time of exposure to [^{3}H]thymidine and the time at which 50% of mitoses are labelled. For instance, if 50% of the mitoses are labelled 2.5 h after removal of [^{3}H]thymidine, one can say that the average duration of the G_2 period is 2.5 h.

As the time interval between removal of [^{3}H]thymidine and fixation of cells increases, the percentage of labelled mitoses increases rapidly to 100%. These are the cells that were in DNA synthesis at the time [^{3}H]thymidine was added. Because of variations in cell cycle length among individual cells, it is not infrequent that the percentage of labelled mitoses does not quite reach 100%. The very shape of the curve of labelled mitoses (7) is an indication that cell cycle times vary among individual cells. The percentage of labelled mitoses remains near 100% for a time which is roughly equivalent to the duration of the phase during which DNA was synthesized and which is called the S-phase. It then drops again and the percentage reaches a low point and, finally, the percentage of labelled mitoses increases again. This second wave of labelled mitoses represents cells that are entering mitosis for the second time since exposure to [^{3}H]thymidine. The determination of the various phases of the cell cycle is based conventionally on the 50% points. Therefore, the 50% points between the two ascending limbs of the curve of percentage labelled mitoses gives the length of the cell cycle, T_c. The interval between the 50% points of the first ascending limb and the first descending limb of the percentage of labelled mitoses curve gives the average duration of the S-phase. We have already said how to measure the G_2 phase and if we now add 45 min of mitosis, the difference between the sum of S + G_2 + M and the total duration of the cell cycle gives the duration of the G_1 period, which is the period between mitosis and S-phase.

The curve of percentage and labelled mitosis is useful if one wishes to determine the effect of growth factors on given phases of the cell cycle. Usually G_1 is the most variable period. For a given cell line S-phase, G_2 and mitosis are fairly constant, although exceptions have been reported to occur.

A very simple method for determining the length of the cell cycle and its phases is by use of flow cytometry. However, in this case, one has to know the length of the cell cycle by some other method; if cell death is negligible, the doubling time of the population is a good approximation of cell cycle length. An illustration of this method is given in *Table 2*, where two populations were considered. The cell cycle distribution was essentially the same in the two populations, but the doubling time was quite different, and therefore the length of each cell cycle phase was also different (cell death was negligible during the observation period in both populations). If one had looked only at the flow cytometry data, one would have come to a totally erroneous conclusion.

5.4 Short method for growth fraction and cell cycle phases

A second method is continuous labelling in tissue cultures, or repeated injections of [^{3}H]thymidine in the living animal. Let us take as an illustration the tissue culture system.

(i) Add [^{3}H]thymidine, 0.02 μCi/ml to exponentially growing cells.
(ii) Fix the coverslips at various intervals after addition of [^{3}H]thymidine (please note that in this case the [^{3}H]thymidine is not removed).
(iii) Determine the fraction of labelled cells as usual.

The fraction of cells labelled within 30 min after exposure of the cells to [^{3}H]thymidine essentially gives the fraction of the cell cycle occupied by the S-phase. If, for instance, the fraction of cells labelled was 40, we would have a simple equation:

$$40 = 100 \times T_s/T_c \qquad [1]$$

Table 2. Illustration of the calculation of duration of cell cycle phases

A	$\%G_1$	%S	$\%G_2$+M
Wild type (w)			
Low	73	14	13
Medium	75	14	11
High	73	14	13

B: Cell cycle parameters of W and R^- cells

	G_1	S	G_2+M	T_c
W (wild type)	33	6	5	44
R^-	73	16	20	109

R^- cells are mouse embryo fibroblasts from knock-out mice, in which the IGF-1 receptor genes have been disrupted by targeted homologous recombination. W cells are from wild type littermates. Cells were seeded at three densities: low ($4 \times 10^3/cm^2$), medium ($8 \times 10^3/cm^2$) and high ($1.6 \times 10^4/cm^2$). Cell growth was measured by determining cell number on a daily basis for three days at each cell density. All counts were done in duplicate. Flow cytometry was performed on each density every day. The percentages shown are from the second day. Average determination for each cycle throughout the experiment were : $G_1 = 72\% \pm 1.7$; $S = 15 \pm 2.1$; $G_2 = 12.5 \pm 2.3$ for wild type, and $G_1 = 66.5 \pm 0.9$; $S = 14.5 \pm 1.3$; $G_2 = 18.5 \pm 2.5$ for R^- cells. The doubling time for each cell type determined during this experiment was 43.6 h for wild type and 109 h for R^- cells.

The duration of the cell cycle phases can then be calculated directly from panel A.

where T_c is the time taken by S-phase, and T_c is the total time taken by the cell cycle.

Under the conditions of cumulative labelling, all cells entering the DNA synthetic phase become labelled, and the time at which 100% of the cells are labelled corresponds to the sum of $G_2 + M + G_1$; that is, $TG_2 + T_M + T_{G_1} = T_c - T_s$. Let us say that the time of 100% labelling was 12 h. The only problem left is to solve a set of simultaneous equations; that is,

$$40T_c = 100 \times T_s \quad [2]$$

$$T_c = 12 \times T_s \quad [3]$$

by solving, $T_s = 8$ h and $T_c = 20$ h. This method is very simple and though it does not separate the duration of G_1, G_2 and mitosis, it does give simultaneously the growth fraction, the length of the cell cycle and the length of the S-phase.

5.5 Flow of cells through the cell cycle

At times it is desirable to determine whether there is a block or a delay in a particular phase of the cell cycle. This may be caused by a drug, by treatment with inhibitory factors or, for instance, in temperature-sensitive mutants, by a defect in one of the gene products that are necessary for cell cycle progression.

A block in S-phase, due to direct inhibition of DNA synthesis, can be detected by exposing the cells at various times after the treatment, to a 30-min pulse of [^{3}H]thymidine. An inhibition of DNA synthesis will result in a quick decrease in the fraction of cells that can be labelled by the pulse exposure, usually within 30 min of treatment.

If, instead, the decrease in the fraction of cells that are labelled by pulse exposure to [^{3}H]thymidine is delayed, then one should suspect a block in G_1, or even later. For instance, suppose that a given drug acts at a point in G_1 which is roughly located 4 h before the S-phase. If that drug is given and the cells are then pulsed with [^{3}H]thymidine at various intervals after treatment, what one will observe is that the percentage of labelled cells (by a pulse of radioactive thymidine) will remain constant, as in controls, for 4 h. Since the cells that were upstream of the block are now inhibited and cannot enter S-phase, the fraction of labelled cells will start decreasing as the cells that were in S-phase are exiting into G_2, but the cells that were located 4 h before the beginning of S-phase are not entering it. By looking at the time required for the decrease in the fraction of labelled cells one can locate the block in the cell cycle. If the block is in mitosis, one would, of course, see a marked increase in the number of mitotic cells, all of which remain unlabelled. Similar experiments can be devised to determine whether there is a block in the flow of cells from S to G_2, or from G_2 to M.

From the comments given above, one ought to be able to avoid the common pitfall of stating that a certain treatment inhibits DNA synthesis, simply because the fraction of labelled cells decreases. This could be due to any block in the cell cycle, that prevents cells from entering S phase. Although DNA synthesis *is* inhibited, it would be an indirect effect, caused by a block in cell cycle progression. Interestingly, the same mistake is made when gene expression is investigated during the cell cycle. We read for instance that a certain treatment inhibits the expression of DNA synthesis genes or G_2 cyclins and cyclin-dependent kinases. This inhibition could be totally indirect, if the block occurs, for instance, in early G_1. This is not splitting hairs; precision of language makes science much easier to understand.

6. Quiescence of cells

A few words on this subject are in order as there is a lot of confusion about the meaning of quiescence, the definition of G_0, etc.

Whether one likes it or not, there is a physiological state of the cells in which the cells do not go through the cell cycle, yet are capable of doing so if an appropriate stimulus is given. The difference between G_0 and G_1 is now firmly established by the fact that in G_0 certain growth-regulated genes are not expressed. The ones that are commonly used are c-*fos* and c-*myc*, but several other genes have been described that are induced when G_0 cells are stimulated to proliferate (10). The fact that gene expression changes, justifies a separation of G_0 from G_1. For the student of the evolution of scientific thought, it may be worthwhile to remember that the existence of a G_0 state was firmly opposed by a large number of investigators for at least 20 years. Nowadays, it has reached that state of acceptance, in which investigators can say that everybody knows that G_0 exists. Indeed, we have now reached the opposite position, and we call G_0 cells those cells that are starved for growth factors, do not grow, but they are not in G_0.

It follows that simply labelling the cultures with [^{3}H]thymidine is not an indication of quiescence. It simply indicates that a population of cells cannot enter S-phase, but it does not really tell us whether those cells are in the G_0 phase in which certain growth-regulated genes are not expressed. A good illustration was reported by Ferrari *et al.* (11) with WI-38 cells, which can be made quiescent simply by contact inhibition. If one takes WI-38 human diploid fibroblasts and plates them in growth medium supplemented with 10% fetal calf serum, the cells grow to confluence. Upon reaching confluence, and without any change of medium, they become quiescent, at least as far as can be judged by labelling with [^{3}H]thymidine. Addition of serum stimulates the re-entry of WI-38 cells into the cell cycle. However, although by day 7 after plating the labelling index is extremely low, Ferrari *et al.* (11) found that certain growth-regulated genes, for instance c-*myc*, are still detectable in Northern blots. The cells are not truly in G_0 and this was demonstrated by the fact that they were still responsive to platelet-poor plasma which is a progression factor. However, if the cells were left confluent for a few more days, up to day 12 after plating, c-*myc*, p53, ornithine decarboxylase, and other growth-regulated genes were no longer detectable and, at this point, the cells were no longer responsive to platelet-poor plasma, although they still responded to serum. These experiments clearly indicate that the labelling index is not a good criterion of quiescence and that to be sure that the cells are in G_0 one has to look at the expression of growth-regulated genes.

References

1. Baserga, R. (1985). *The Biology of Cell Reproduction*. Harvard University Press, Cambridge, MA.
2. Burton, K. (1956). *Biochem. J.*, **62**, 315.
3. Baserga, R. (1984). *Exp. Cell Res.*, **151**, 1.
4. Mercer, W. E., Avignolo, C., Galanti, N., Rose, K. M., Hyland, J. K., Jacob, S. T., and Baserga, R. (1984). *Exp. Cell Res.*, **150**, 118.
5. Sisken, J. E., Bonner, S. V., Grasch, C. D., Powell, D. E., and Donaldson, E. S. (1985). *Cell Tissue Kinet.*, **18**, 137.
6. Zieve, G. W., Turnbull, D., Mullins, J. M., and McIntosh, J. R. (1980). *Exp. Cell Res.*, **126**, 397.

7. Baserga, R. and Malamud, D. (1969). *Autoradiography*. Harper & Row, New York.
8. Mendelsohn, M. L. (1962). *J. Natl. Cancer Inst.*, **28**, 1015.
9. Stanners, C. P. and Till, J. E. (1960). *Biochim. Biophys. Acta*, **37**, 406.
10. Kaczmarek, L. (1986). *Lab. Invest.*, **54**, 365.
11. Ferrari, S., Calabretta, B., Battini, R., Cosenza, S. D., Owen, T. A., Soprano, K. J., and Baserga, R. (1988). *Exp. Cell Res.*, **174**, 25.

2

Cell cycle analysis by flow cytometry

HARRY A. CRISSMAN

1. Introduction

Mechanisms that control and regulate the cell cycle are currently being intensively studied in numerous laboratories. Central to these studies is the basic understanding of those regulatory mechanisms that appear to be lost or dysfunctional when cells become transformed from the normal to the malignant state. This information can lead to a better understanding of the nature of cancer and potentially result in the design of more effective chemotherapeutic agents or even to the intervention of those processes that lead to the development of cancer. Studies in these areas will necessarily rely on the constant assessment of cell cycle distribution and cycle progression.

For more than 25 years, quantitative fluorescent staining and flow cytometric (FCM) analysis of cellular DNA content has remained one of the most rapid and reliable approaches for obtaining cell cycle frequency distributions of cell populations. The precision in these studies relies on the specificity of the staining methods for tagging DNA in cells and the efficiency of flow instruments for quantitative fluorescence analysis. Early studies, comparing computer-fit analysis of DNA content histograms to data obtained by conventional tritiated-thymidine labelling and autoradiography, confirmed the accuracy of the FCM technique for cell cycle analysis.

The ease and rapidity of FCM analyses provides advantages such as the abilities to (a) access the cell cycle distributions during a study, and alter the experiment if desired; (b) monitor the various phases in populations composed of slowly progressing or arrested cells; (c) detect abnormalities in progression through mitosis such as non-dysjunction or polyploidization and (d) detect aneuploid subclones within a cell population to provide information of clinical prognostic value. In addition, cells can be sorted based on DNA content, so there is not the need to synchronize cells for phase-specific biochemical analysis.

The limitation of DNA content–cell cycle frequency analysis is that the method provides no information on the cycling capacity and the rate at which cells traverse the cell cycle. The cell cycle frequency histogram provides a static image of the cell population showing only the position of all cells at a given moment, but the proportions of cycling and non-cycling cells within the various phases are not known. Sequential sampling of the population during an experiment does alleviate this problem to some extent.

The development of multiparameter flow cytometry has provided for analysis of other physiological parameters, which, in addition to DNA metabolism, regulate and control cell proliferation. Many current studies involve labelling and measuring cellular constituents such as proteins and RNA simultaneously with DNA. Cellular levels of such descriptors, and others, are known to be important indicators of cell cycle progression capacity, cell growth and function. In addition, bromodeoxyuridine (BrdUrd) incorporation studies have refined methods for analysis of cell cycle progression, and the use of monoclonal antibodies in FCM immunofluorescence studies have improved techniques for examining and quantitating proliferation markers in different phases of the cell cycle. In this chapter some of the various techniques will be described together with preparative and analytical methods.

2. Cell preparation and fixation

The choice of preparation and fixation is determined by the biological sample and the constituents to be fluorochrome-labelled and analysed. Since DNA measurements are often coupled with analyses of other cellular constituents, preparation and fixation must ensure that all the properties of interest are optimally preserved. For analysis of DNA content, only intact nuclei need be preserved, whereas labelling of DNA and cytoplasmic or cell membrane components, requires mild dispersal methods that provide intact cells.

Membranes of viable cells exclude many of the fluorochromes but can be permeabilized by brief treatment with non-ionic detergents, or with hypotonic solutions, or proteolytic enzymes for rapid DNA staining. However, membrane components and cytoplasmic constituents are lost by these treatments. Alternatively, ethanol fixation perforates cell membranes, but preserves most cytoplasmic materials and does not appear seriously to affect dye binding to DNA.

Protocol 1. Ethanol fixation of cells

1. Harvest cells from culture medium or from tissue dispersal solutions by centrifugation.
2. Perform cell count, tally total number of cells and calculate the volume of fixative required to yield 10^6 cells/ml.
3. *Thoroughly* resuspend the cells in one part cold 'Saline GM' (g/litre: glucose 1.1; NaCl 8.0; KCl 0.4; $Na_2HPO_4 \cdot 12\ H_2O$ 0.39; KH_2PO_4 0.15)

containing 0.5 mM EDTA for chelating free calcium and magnesium ions.

4. Add three parts cold, 95% non-denatured ethanol to the cell suspension with mixing to produce a final ethanol concentration about 70%.

Protocol 2. Nuclear preparation

A protocol described by Gurley *et al.* (1) has proven useful for preparing CHO and HL-60 nuclei.

1. Harvest cells from culture medium by centrifugation at 4°C for 8 min at 200 *g*, and remove the supernatant by aspiration.
2. Wash cells twice with 5.0 ml of cold Saline GM (see *Protocol 1*).
3. Add 4.0 ml of RSB swelling solution, containing 0.01 M NaCl, 0.0015 M $MgCl_2$, 0.01 M Tris (pH 7.4), and 50 μg/ml RNase (Worthington Biochemical Corporation) to the cell pellet.
4. Vortex vigorously for 30 sec and allow to stand on ice for 5 min.
5. Add of 0.5 ml of 10% Nonidet P-40 (NP-40) detergent (Polysciences Inc.) to the sample and vortex again vigorously for 30 sec and put on ice for 15 min.
6. Add 0.5 ml of 5% sodium deoxycholate (DOC) detergent (Mann Research Laboratories Inc.) to the cell sample, vortex for 30 sec and then put on ice for another 15 min.

Hedley *et al.* (2) and Hedley (3) devised methods for preparing nuclei from formalin-fixed, paraffin-embedded tissue. Following removal of paraffin with xylene or the less toxic commercial product 'Histoclear', tissue samples are rehydrated, treated for 30 min with aqueous 0.5% pepsin–HCl and subsequently stained for DNA with 4′,6-diamidino-2-phenylinidole (DAPI) or propidium iodide (PI).

3. DNA-specific fluorochromes commonly used for staining

A significant number of fluorochromes are available with different spectral properties and/or modes of binding to DNA. Staining protocol using the different DNA-specific fluorochromes are equally efficient for FCM-cell cycle analysis, and the choice of a particular technique and/or fluorochrome depends primarily on the excitation wavelengths available and the spectral properties of other fluorochromes to be used in combination with the DNA stain.

3.1 DNA reactive Hoechst dyes, DAPI and DIPI

The Hoechst dyes are non-intercalating, benzimidazole derivatives that bind preferentially to AT base regions and emit blue fluorescence when excited by UV light at about 350 nm. Latt (4) showed that the fluorescence of Hoechst 33258 was quenched when bound to BrdUrd-substituted DNA, and Arndt-Jovin and Jovin (5) first demonstrated the use of both Hoechst 33258 and 33342 for quantitative staining and sorting of viable cells. The structural and spectral properties of DAPI and DIPI are similar to the Hoechst dyes and either can be used for DNA staining in ethanol-fixed cells. However, neither DAPI nor DIPI are as sensitive to BrdUrd as the Hoechst dyes.

3.2 Mithramycin, chromomycin, and olivomycin

Mithramycin (MI) is a green–yellow fluorescent, DNA-reactive antibiotic similar in structure and dye binding characteristics to chromomycin A3 (ChA3) and olivomycin. These compounds, complexed with magnesium ions, preferentially bind to GC base regions by non-intercalating mechanisms. Spectrofluorometric analysis of mithramycin or chromomycin A3–Mg complexes bound to DNA in PBS shows two excitation peaks, a minor peak in the UV (about 320 nm) and a major peak at about 445 nm. A broad green–yellow emission spectrum is observed with a peak at 575 nm (6). Olivomycin by comparison has slightly lower wavelength excitation and emission characteristics.

3.3 Propidium iodide (PI), ethidium bromide (EB), 7-amino-actinomycin D, and hydroethidine (HE)

The red fluorochromes, propidium iodide (PI) and ethidium bromide (EB), have similar chemical structures and both intercalate between base pairs of double-stranded DNA and RNA without base specificity. DNA-specificity requires pre-treatment of fixed cells with RNase. Spectral studies on PI bound to calf thymus DNA in PBS show two excitation peaks, a minor but substantial peak in the UV region (340 nm) and a major peak at about 540 nm. One emission peak was observed at 615 nm.

Neither PI nor EB penetrate viable cells, but cells with damaged membranes stain readily, so these dyes are often used for differentiating and quantitating viable and dead cells in a given population. Methods employing procedures used for permeabilizing cell membranes and/or tissue disaggregation (7, 8) have been developed for rapid staining of unfixed cells with PI.

Another DNA-specific dye 7-amino-actinomycin D (7-Act D), is a red fluorescent analogue of actinomycin D (excitation 540 nm), that intercalates into GC regions in DNA. Zelenin *et al.* (9) performed a detailed study on the use of 7-Act D for cell staining and FCM analysis.

Hydroethidine (HE) is a fluorescent compound produced by the reduction of EB. Gallop *et al.* (10) showed that HE rapidly enters viable cells, where it is enzymatically dehydrogenated to ethidium which then intercalates into double-stranded DNA and RNA and fluoresces red. Non-reacted HE in the cytoplasm fluoresces blue when excited at 370 nm. Using 535 nm excitation only red (ethidium) fluorescence is seen.

3.4 Acridine Orange

Acridine Orange is a metachromatic dye that fluoresces green while intercalated between DNA base pairs and red when stacked on RNA. The chemistry, binding properties, and flow cytometric applications of the dye have been carefully examined and experimentally exploited by Darzynkiewicz (11).

3.5 TOTO-1, YOYO-1, TO-PRO-1, and YO-PRO-1

TOTO-1 and YOYO-1 are modified dimers of the dyes Thiazole Orange and Oxazole Yellow, respectively, with DNA and RNA specificity (Molecular Probes Inc.) and that bind to DNA by intercalation. The relative fluorescence intensities of these dyes bound to the DNA bands in gels indicate that dye-binding is proportional to DNA content rather than to base composition. Haugland (Molecular Probes, Inc., Handbook 1992–94) has reported quantum efficiencies (QE) of 0.34 for TOTO-1, and 0.52 for YOYO-1. The ultrasensitivities of TOTO and YOYO, coupled with their excitation characteristics, 488 and 457 nm wavelengths, respectively, have made these fluorochromes useful for flow cytometry (FCM). Previous FCM studies demonstrated DNA content analyses of RNase-treated nuclei and RNase-treated fixed cells, stained with μM concentrations of TOTO and YOYO and their respective monomers, TO-PRO-1 and YO-PRO-1 (12).

4. Cell staining

4.1 Viable cell staining with Hoechst 33342

Since synchronization protocol may potentially perturb the cell cycle, DNA-specific staining in viable cells coupled with FCM analysis and sorting, provides an alternative approach for selecting and recovering cells from various phases of the cell cycle. The sorted cells can be cultured and examined with regard to long-term viability, functional activity, as well as immunological and other physiological properties.

However, some precautions should be noted since dyes such as HO 33342, that bind to DNA can potentially interfere with DNA replication and such dyes have been shown, under some conditions, to impair viability (13). However, two recent studies showed that treatment of viable cells with membrane-interacting agents can improve Hoechst uptake and improve survival in some cell types.

We showed that in certain cell types the membrane-potential modifying mitochondrial stain, DiO-C5-3, when applied to viable cells in conjunction with Hoechst 33342, increased cellular uptake or retention of the Hoechst dye twofold and provided coefficient of variation (CV) values of about 3.0%

compared with 8.3% for cells treated with Hoechst alone (14). Krishan (15) demonstrated that calcium channel blocking agents, such as verapamil, can also increase Hoechst stainability in viable cells that are normally refractory to Hoechst uptake. Those studies indicated that rapid metabolic dye efflux often accounts for poor Hoechst stainability.

However, even with optimal staining conditions, dye uptake, cytotoxicity, DNA binding and analytical resolution, as judged by coefficients of variation (CV) in the FCM–DNA histograms are cell type dependent. Cells with damaged membranes or cells with membranes perforated with non-ionic detergents (i.e., NP-40, Triton X100) stain rapidly. For more details see Crissman *et al.*(14).

Protocol 3. Viable cell staining for DNA content

1. Add HO 33342 (final concentration 2.0–5.0 μg/ml) to cells in culture medium (37°C) for incubation periods of 30–90 min depending on the cell type.
2. Remove and place aside medium containing HO 33342.
3. Trypsinize cells from monolayer or harvest cells directly from suspension culture.
4. Neutralize trypsin with medium set aside containing Hoechst and perform cell count.
5. Centrifuge cells and resuspend cells in HO 33342 containing medium to a density of about 7×10^5 cells/ml for analysis.

4.2 Staining of fixed cells and unfixed nuclei

Protocol 4. Staining of ethanol-fixed cells

1. Centrifuge fixed cells and aspirate the fixative.
2. Resuspend cells in the appropriate stain solution[a] at a density of about 8×10^5 cells/ml and vortex cell suspension.
3. Analyse stained cells after about 1 h at room temperature.

[a] DNA fluorochromes are available from several commercial companies including: Molecular Probes, Inc., Polysciences, Inc., Calbiochem-Behring Corp., Sigma Chemical Co. Mithramycin is available from Pfizer Co. Stock solutions of most dyes are prepared in PBS at concentrations of 1.0 mg/ml; however, DAPI and the Hoechst dyes should be prepared in distilled water, since at relatively high concentrations these dyes tend to precipitate in PBS. Stock solutions, refrigerated in dark-coloured containers or wrapped in foil, have been used for at least one month without noticeable degradation. Stain solutions are prepared in PBS using the fluorochrome of choice at the following concentrations: PI, EB or 7-Act D 15 μg/ml; MI or ChA3, 50 μg/ml with 20 mM $MgCl_2$; DAPI, 1–2 μg/ml and Hoechst dyes at 0.5–1.0 μg/ml. RNase (50–100 μg/ml) is added to stain solutions containing PI or EB. Dye concentrations required to obtain optimal results may vary slightly depending on the flow instrument used.

Protocol 5. Staining of unfixed nuclei

1. Add dyes[a] directly to the nuclei preparation without centrifugation, which tends to cause clumping of the nuclei.
2. Stain for at least 1 h at room temperature prior to analysis.

[a] Stock solutions of TOTO, YOYO, TO-PRO-1 and YO-PRO-1 (1.0 mM) (Molecular Probes, Inc.) are added directly to fresh nuclei in the isolation buffer, to final dye concentrations of 4.0×10^{-6} M for TOTO and YOYO and 4.0×10^{-5} M for TO-PRO-1 and YO-PRO-1. Fluorescence of nuclei remains stable for at least 5–6 h.

5. Instrumention setup and analysis

Viable cells in suspension stained with HO are analysed in equilibrium with the dye, using a UV laser beam (~350 nm) or a mercury-arc lamp excitation source and analysing emission above 400 nm. For sorting viable cells the laser excitation power should be set as low as possible to reduce potential phototoxicity to the HO-stained cells. The CV in the DNA histogram will increase somewhat at the lower power, but it is necessary to compromise resolution for cell survival. When using MI, chromomycin A3 or YOYO-1 staining, cells are excited at 457.9 nm and fluorescence analysed above 500 nm. Cells stained with PI, EB, TOTO-1 or 7-Act D are usually excited at 488 nm and emission collected above 515 nm.

Stained samples should be allowed to run in the FCM for about 1–2 min until the dye solution and the sheath fluid are equilibrated. The time required for stabilization will vary from instrument to instrument, depending on the hydrodynamics of the flow system.

Experimental drugs which interact directly with DNA or interfere with DNA metabolism can have effects on subsequent staining. In *in vivo* studies, Alabaster *et al.* (16) found differences in MI-DNA stainability of untreated (control) L1210 ascites cells and cells treated with a combination of cytosine arabinoside and adriamycin. Krishan *et al.* (17) found that adriamycin diminished PI staining and/or fluorescence.

All DNA-specific compounds should be considered potential carcinogens and therefore handled with some precautions. Solutions should be immediately washed from the skin if contact is encountered. Aerosols created during long-term cell sorting contain some of the stain solution and care must be taken to avoid inhaling the dyes.

5.1 Standards

Stained, nucleated trout or chicken red blood cells (RBC) have been used as internal markers for determining the DNA index (DI) (18), as calculated by dividing the G_1 peak channel of an aneuploid population by the G_1 peak of normal cells analysed at the same instrument gain setting. Fluorescent microspheres are often used to align the excitation beam on the cell stream and to calibrate the linearity of the instrument gain setting. Linearity is achieved when the increase in the relative intensity of the microspheres is proportional to the electronic gain setting.

5.2 Application

When ethanol-fixed cells are stained with mithramycin (*Protocol 4*) and analysed by flow cytometry results are obtained as shown in *Figure 1*. The DNA content histograms shown are for a control (untreated) human skin fibroblast (HSF) cells (A) and for a HSF cell population treated for 8 h with a 100 nM concentration of the protein kinase inhibitor, staurosporine (STSP) (B). The 8 h treatment with STSP can be seen to slightly decrease cycle progression through the S and block cells in G_2/M phases of the cell cycle. Treatment for 18 h will arrest HSF cells in the G_1 and G_2 phases of the cell cycle (data not shown) (19).

6. Simultaneous DNA and protein assay

Correlation in the synthesis and accumulation of DNA and protein plays an important role in regulating cycle traverse capacity, cell division, growth and size. Consistency in both the cycle generation time and the cellular protein content distribution, as reflected by the regularity of the volume distribution of the population, are controlled by transcriptional and translational processes that rigidly couple temporal metabolism of these macromolecules.

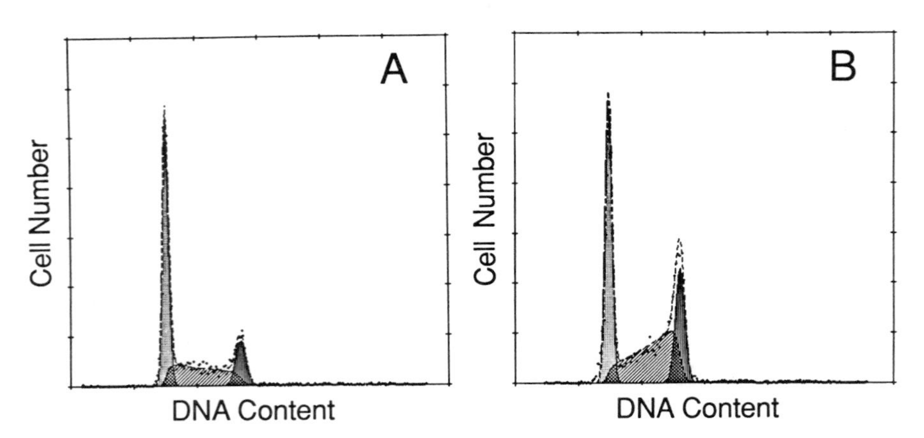

Figure 1. DNA content histograms and computer-fit displays for asynchronous human skin fibroblasts cells (A) or cells treated with staurosporine (B) as described. Cells were fixed in ethanol and stained with mithramycin prior to analysis by flow cytometry. The percentages of cells in G_0/G_1, S and G_2/M phases, respectively, are: 55, 29 and 16% for the untreated cells (A), and 37, 41 and 24% for the staurosporine-treated cell population (B).

Although DNA content increases only during S phase, protein synthesis is constant, and increases linearly across the cell cycle of exponentially growing mammalian populations.

Flow cytometry (FCM) provides for biochemical measurements of both DNA and protein in single cells thus allowing for subsequent correlation of these metabolic parameters in studies on the relationship between cell growth and the cell division cycle. Simultaneous measurement of DNA and protein also allows assessment of the ratio of protein to DNA of cells located at particular stages of the cell cycle under a variety of experimental conditions. Two-colour staining of DNA and protein can be most easily accomplished using a combination of propidium iodide (PI) and fluorescein isothiocyanate (FITC) (20) or by staining with a combination of DAPI and sulpharhodamine 101 (21), that only requires UV excitation and blue and red fluorescence analysis of DNA and protein content, respectively. A combination of PI–FITC has also been used to examine and compare directly nuclear protein and DNA contents of cells in various phases of the cell cycle (22, 23).

The one-step staining procedures involving several dye combinations have been applied to a variety of cell types. By controlling the dye concentrations it is possible to minimize the dye–dye interactions as well as the excessive background fluorescence (24).

6.1 Cell staining

Protocol 6. Cell staining

1. Centrifuge ethanol-fixed cells prepared by *Protocol 1* and aspirate fixative.
2. Resuspend cells in appropriate dye combination for staining.[a]
3. Allow sample to stand for at least 1 h at room temperature prior to analysis.

[a] *PI–FITC–RNase staining*: PI (15 μg/ml), FITC (0.05–0.10 μg/ml) prepared in PBS containing 50 μg/ml RNase.
DAPI-SR101 staining: DAPI (2 μg/ml) and SR101 (15–20 μg/ml) in PBS.
Stock solutions of reagents are prepared at 1.0 mg/ml and stored refrigerated for at least one month unless indicated otherwise. Propidium iodide (Molecular Probes, Inc.) is prepared in PBS, and DAPI (Molecular Probes, Inc.) is prepared in distilled water. Fluorescein isothiocyanate ((FITC), isomer 1, BBL, Division of Becton Dickinson Co.) is prepared in absolute ethanol just prior to use. Sulpharhodamine 101 (Molecular Probes, Inc.) is dissolved in dimethyl sulphoxide (DMSO). Stock solutions of RNase (Worthington) are prepared in PBS.

6.2 Instrument setup and analysis

6.2.1 PI–FITC–RNase cell samples

Single laser excitation at 488 nm is used for analysis of the stained cell sample. The filter combination consists of a 600 nm long pass dichroic filter, with

515 nm long pass and 590 nm long pass filters for detection of FITC and PI, respectively. In some cases, fluorescence compensation may be necessary to adjust for overlap in the DNA and protein fluorescence. Subsequent bivariate analysis is used to correlate the individual parameters.

6.2.2 DAPI-SR101-stained cell samples

A single UV, laser or mercury arc source, can be used to excite both of these dyes simultaneously. The filters used were a 500 nm long pass dichroic, with 400 nm long pass and 610 long pass filters for analysis of blue and red fluorescence, respectively.

6.3 Applications

Simultaneous analysis of DNA–protein contents have been useful for making nuclear-to-cytoplasmic ratio determinations (25). As cells traverse the laser beam, the time duration of the DNA and protein fluorescence signals provide a proportional measurement of the nuclear and cytoplasmic size, respectively. Also gated analysis of the single parameter DNA histogram can provide the protein content histograms and mean values for cells in various phases of the cell cycle (20). Considering the ease and simplicity of DNA–protein determinations and the increased amount of information available for gauging experimental effects, it would seem worthwhile in the design of a particular study to include the addition of protein content measurements along with the routine DNA–cell cycle analyses.

Nuclear to cytoplasmic ratio analysis requires the use of a FCM system in which the laser beam is focused to a thin elliptical slit that is smaller than the diameter of the nucleus (25). The ratio of the time duration of the red to green signals provides the relative nuclear to cytoplasmic size relationship on a cell by cell basis. Analysis of PI–FITC stained CHO cells (*Figure 2*) shows a good correlation in nuclear to cytoplasmic relationship throughout the cell cycle (*Figure 2F*) as would be expected for an exponentially growing, homogeneous cell population.

7. Cellular bromodeoxyuridine and DNA contents

Analysis of DNA content by flow cytometry (FCM), coupled with cellular BrdUrd incorporation data provides phase-specific kinetic information that allows further subdivision of the cell cycle based on cycle traverse capacity. The procedure provides the cycle position of cells as well as the relative percentage of cells traversing from one phase to the next. Information from such studies has been useful for studying the differences in the kinetic patterns of normal and tumour cells, and also for investigating the effects of cycle perturbing agents on cell cycle progression.

Probably the most widely used assay is that devised by Dolbeare *et al.* (26) that combines propidium iodide (PI) DNA staining with immunofluorescent-

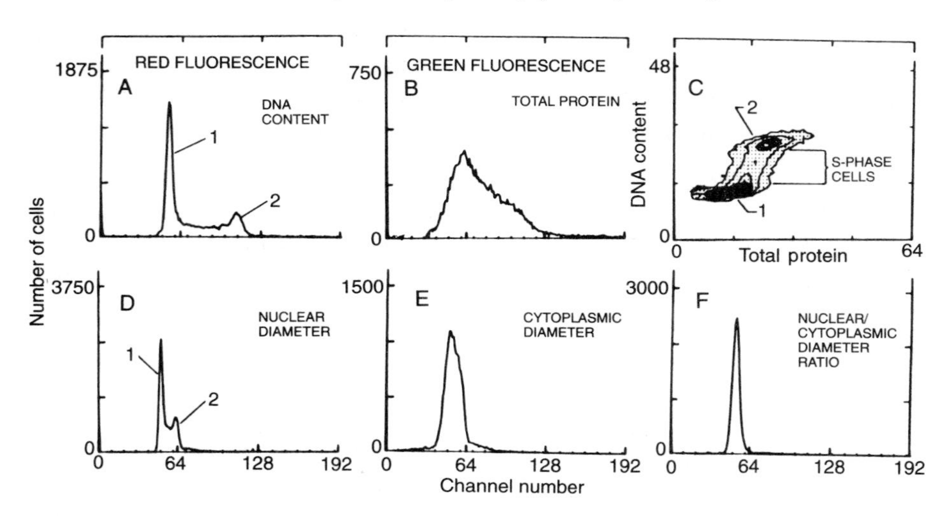

Figure 2. Frequency distribution histograms of line CHO cells fixed in 70% ethanol and stained with propidium iodide (PI) (DNA content) and fluorescein isothiocyanate (FITC) (total protein): (A) DNA content (red fluorescence); (B) total protein (green fluorescence); (C) DNA vs protein two-parameter contour diagram; (D) nuclear diameter; (E) cytoplasmic diameter; and (F) nuclear-to-cytoplasmic diameter ratio distribution. The *x*-axes (channel number) are proportional to: red fluorescence (A); green fluorescence (B); red fluorescence signal time duration (nuclear diameter) (D); green fluorescence signal time duration (cytoplasmic diameter) (E); and the ratio of red-to-green fluorescence signal time duration (F). Cells in G_0/G_1 and cells in G_2/M are denoted by 1 and 2, respectively.

labelling of BrdUrd as described by Gratzner (27). This method requires partial denaturation of cellular DNA by heat or acid treatment to expose the incorporated BrdUrd for antibody binding; however, the technique has the advantage that it requires only a single laser (488 nm) for excitation of both PI and fluorescein-labelled BrdUrd.

Other FCM procedures for DNA–BrdUrd analysis utilize a cytochemical method involving analysis of quenching of the fluorescence of Hoechst 33258 when bound to A-BrdUrd regions in double stranded DNA (4) Some techniques employ a second DNA fluorochrome, such as ethidium bromide (EB), that is not significantly affected by BrdUrd in combination with Hoechst for cell staining, and analysis performed using a single UV excitation source (28–32). The technique has been applied primarily for continuous BrdUrd labelling studies, since labelling periods of 4–6 h or longer are required for cells to incorporate sufficient amounts of the base analogue for detection by HO quenching.

Another FCM method for the analysis of cycle progression employs a two-colour technique that can potentially detect BrdUrd-labelled DNA in cells treated for 30 min or less (33). Staining incorporates two non-intercalating,

DNA fluorochromes: Hoechst 33342 (HO) and GC binding mithramycin (MI), a dye whose stoichiometry for DNA content is not affected by BrdUrd. Using dual-laser excitation, for performing differential fluorescence analysis on the two fluorochromes, a measure of the cellular BrdUrd content as well as the MI–DNA content fluorescence is obtained. The technique requires only one-step staining; it is mild, and therefore, cell loss and loss of other important cellular markers such as RNA, and proteins, including cellular antigens are minimal.

7.1 Differential fluorescence analysis of BrdUrd content

Essentialy, the technique represents a quantitative approach for measuring a small decrease in Hoechst 33342 (HO) fluorescent intensity when the dye is bound to BrdUrd incorporated into DNA during short pulse-labelling periods. Quantitative FCM measurement of the HO quenching is accomplished by electronic cell-by-cell subtraction of the HO fluorescent signal from the signal of a second dye, such as the GC binding stain, mithramycin (MI), that is not affected by BrdUrd. Also HO and MI have different spectral and DNA binding properties and neither dye–dye interference nor energy transfer between the MI and HO significantly affects FCM BrdUrd analysis. Both HO and MI bind stoichiometrically to DNA content in non-BrdUrd labelled cells. However, in a cell population that has been pulse-labelled with BrdUrd (i.e., only S phase cells are labelled), differential fluorescence values near zero are obtained for the MI minus HO fluorescent signal differences for the G_1 and G_2+M cells in which the fluorescent intensity of both dyes remains proportional to DNA content. Cells in S phase containing BrdUrd, produce HO fluorescent signals that are reduced in proportion to the quantity of BrdUrd-substituted DNA, whereas the MI fluorescent intensity in these same cells still remains proportional to the DNA content. Differential fluorescence analysis of S phase cells will then yield MI minus HO differences that are greater than zero and proportional to BrdUrd content.

7.2 BrdUrd pulse labelling

For pulse-labelling studies, CHO cells are routinely treated in culture with 30 μM BrdUrd for 1 h at 37°C. Cultures are not exposed to strong light over this period since the BrdUrd is light sensitive and can deteriorate. Also, the DNA of cells incorporating BrdUrd can be damaged when exposed to strong light and thus fail to incorporate sufficient quantities of BrdUrd for FCM detection.

The concentration of BrdUrd and the length of the labelling period can vary depending on the cell cycle time and the rate of DNA synthesis. Rapidly growing cells, for example, may require only a 30-min pulse, but more slow-growing cells may require pulses of 1–2 h before they have incorporated detectable quantities of BrdUrd. The pulse duration as well as the BrdUrd concentration must be determined empirically. Excessively high concentrations of BrdUrd can be toxic to cells, and comparative growth studies and FCM analyses of BrdUrd-treated and untreated cell populations are necessary to determine optimal conditions for BrdUrd labelling.

Protocol 7. BrdUrd pulse labelling

1. Add appropriate concentration of sterile BrdUrd (10–30 μM) directly to the culture medium of cells and incubate cells in the dark at 37°C for desired time period.[a]
2. Harvest cells from monolayer with trypsin or directly from suspension culture and place on ice.
3. Centrifuge cell suspension, aspirate medium and fix in 70% ethanol using *Protocol 1*.

[a] Stock solutions of BrdUrd (5-bromodeoxyuridine, Sigma Chemical Co.) are freshly prepared in distilled water (1.0 mg/ml) and sterilized by filtration. The BrdUrd solution should be refrigerated in vials wrapped in foil to protect it from light.

7.3 Cell staining with Hoechst 33342 (HO) and mithramycin (MI)

Staining experiments are performed initially on untreated cells to determine the lowest concentrations of HO and MI required to obtain DNA content histograms of good quality with low coefficient of variation (CV) values. Ideally, both the HO–DNA content and the MI–DNA content histograms of non-BrdUrd labelled cells should yield the same percentage of cells in the various phases of the cell cycle. Under these conditions the dye–dye interactions will be minimal and the subtraction of the HO fluorescent signal from the MI signal in unlabelled cells, on a cell by cell basis, will be as close to the ideal value of zero expected for unlabelled cells. Dye concentrations listed have been used for a variety of cell types and are suggested for initial staining trials.

Protocol 8. Cell staining with Hoechst 33342 (HO) and mithramycin (MI)

1. Centrifuge ethanol-fixed cells prepared as in *Protocol 1* and aspirate fixative.
2. Resuspend cells in dye solution containing HO 33342 (0.5 μg/ml) and MI (5.0 μ/ml) in PBS containing 5 mM $MgCl_2$.[a] Cell density is adjusted to about 7.5×10^5 cells/ml.

Protocol 8. *Continued*

3. Analyse cells after 1 h staining at room temperature.

[a] Stock solutions of Hoechst 33342 (Molecular Probes, Inc.) dissolved in distilled water at 1.0 mg/ml, can be stored in the refrigerator for at least one month in a foil-wrapped container. Mithramycin (MI) (Pfizer Co.), 2.5 mg/vial, is dissolved in PBS at 1.0 mg/ml and can be stored in the refrigerator for at least one month. Solutions of $MgCl_2$ (250 mM) are prepared in distilled water and may be stored at room temperature for at least one month.

7.4 Instrument setup for differential fluorescence analysis

A multi-laser, FCM system (34) is used and two lasers are operated, one in the UV (333.6–363.8 nm) and one tuned to 457.9 nm. The laser beams are separated by 250 μm to provide sequential excitation and analysis of each fluorochrome (i.e, HO and MI respectively). The HO fluorescence is measured over the 400–500 nm range, whereas the MI fluorescence is measured above 500 nm. A 500 nm long pass dichroic filter is used for these studies. The electronic signal gains are adjusted so that the G_1 peaks of the FCM generated HO– and MI–DNA content histograms are initially at the same channel number. The fluorescence signals are then subtracted electronically (i.e, MI minus HO) on a cell-by-cell basis using a differential fluorescence analysis method (35). These difference measurements reflect quantitatively the quenching of HO fluorescence, which is directly proportional to cellular BrdUrd content. The arbitrary adjustment of the G_1 peak positions for both dyes in the same channel initially sets the zero value for subtraction process. The initial assignment of the zero value is valid since the G_1 cells in pulse-labelled cells do not contain BrdUrd, therefore do not exhibit HO quenching, and should yield MI minus HO values of zero.

7.5 Analysis of stained cells

Data from a pulse-labelling experiment performed on non-BrdUrd-labelled (A, B, C) and BrdUrd-labelled (D, E, F) Chinese hamster (line CHO) cells is shown in *Figure 3*. The HO 33342 distribution (*y* axis) in *Figure 3D*, compared to *Figure 3A*, shows a curvature in the S phase cell resulting from the BrdUrd/HO fluorescence quenching. A similar effect is seen for labelled S phase cells in the ratio (HO/MI) data (*Figure* 3E), compared to ratio data (*Figure 3B*) of non-BrdUrd treated cells. The BrdUrd content, as derived from the MI minus the HO difference measurements, is represented on the *y* axis in *Figure 3F* while the difference measurement for non-BrdUrd treated cells (*Figure 3C*) shows only the slight differences in stainability of DNA by the two different dyes.

The simplicity, sensitivity and accuracy of BrdUrd analysis by differential fluorescence analysis makes the procedure quite applicable for basic and

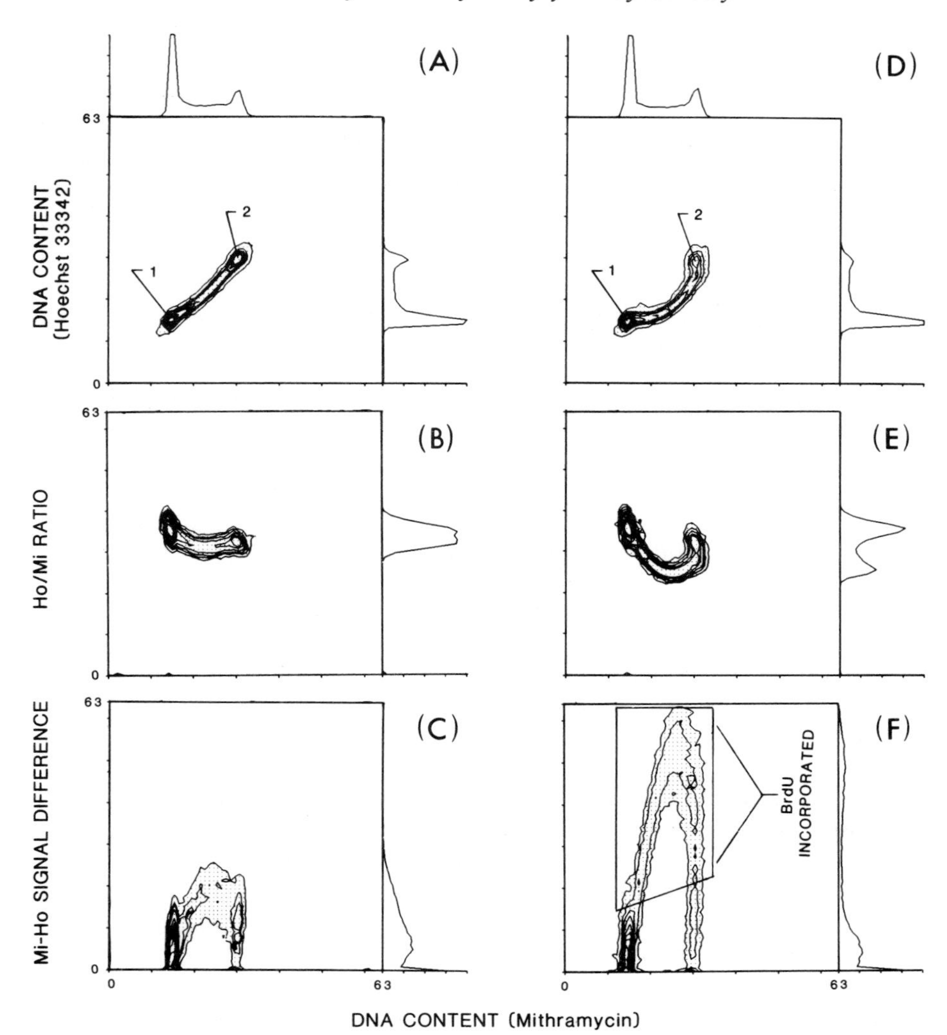

Figure 3. Bivariate contour distributions for untreated CHO cells (A, B, C) and for cells treated with 30 μM BrdUrd for 1 h (D, E, F) prior to ethanol fixation, staining with Hoechst 33342 and mithramycin and dual-laser flow cytometric analysis as described. All the bivariate distributions show MI fluorescence (DNA content) on the *x* axis. In A and D and HO fluorescence is plotted on the *y* axis; in B and E the ratio of HO to MI is on the *y* axis, and in F the *y* axis represents the BrdUrd content as obtained by differential fluorescence analysis of the MI fluorescence signal minus the HO signal. The bivariate in C for the non-BrdUrd treated sample shows only the slight differences in stainability of DNA by the two dyes.

clinical studies. For routine studies, sample preparation should be simple and rapid, and preparation of the samples for analysis free of manipulations that would induce cell loss and variability from sample to sample, due to repetitive sample handling during the preparative procedure. The preparative and analytical procedure described above fulfils these criteria: one-step ethanol fixation, preventing any cell loss; one-step staining with a stain cocktail and differential fluorescence analysis in the stain solution. Although most commercial FCM systems do not have capabilities for differential fluorescence analysis, it is possible to acquire the data in list mode and perform the subtraction of the MI and HO fluorescence signals on a cell-by-cell basis using an appropriate computer program (36). The only significant disadvantage is that this method requires a laser with UV capabilities for exciting HO in order to analyse fluorescence quenching by BrdUrd. Ideally, dyes with properties similar to HO, but with excitation in the visible range, will become available in the near future.

8. Correlated, three-colour, DNA–RNA–protein determinations

Transcriptional and translational processes regulate and couple the temporal metabolism of DNA, RNA, and protein. Flow cytometry (FCM) performs multiple biochemical measurements on single cells, thereby allowing for subsequent correlation of the various metabolic parameters. Simultaneous measurement of DNA, RNA, and protein permits assessment of the ratio of RNA to DNA and RNA to protein per cell, and such information can serve as a sensitive gauge of the metabolism of cells located at particular stages of the cell cycle under a variety of experimental conditions. A FCM method for direct determination and correlation of DNA, RNA, and protein in individual cells (37) involves modification of procedures for fluorochroming DNA with Hoechst 33342 (HO) RNA with pyronin Y (PY), and protein with fluorescein isothiocyanate (FITC). Analysis can be performed in a three-laser FCM system, such as that previously described (34).

A variety of cycle-perturbing agents, including drugs, induce a differential uncoupling of normal synthetic patterns, causing a disproportionate accumulation of these cellular constituents. These conditions may lead to states of unbalanced growth, loss of long-term viability, and eventual cell death. In one study (38), correlated analysis of DNA, RNA, and protein by FCM was used to determine the effects of adriamycin (AdR) on exponentially growing CHO cells. AdR treatment induced a differential response in metabolism of DNA, RNA, and protein in the cell population. At 15 h after a 2 h drug treatment period, 85% of the cells were arrested in the G_2 phase, and this subpopulation had mean ratio values for RNA to DNA and RNA to protein that were 44% and 31% elevated, respectively, above ratio values of control cells. These data indicated that the G_2-arrested, AdR-treated cells were in a gross state of unbalanced growth. However, the cells remaining in G_1 phase (i.e., 15%) had RNA to DNA and RNA to protein ratio values almost identical to control values. Survival studies showed that compared with the exponential control CHO cell population the AdR-treated exponential population had only a 12% surviving fraction. Subsequent viable cell sorting studies (39), based on bivariate DNA and cell volume analysis, and plating efficiency assays confirmed the accuracy of the DNA, RNA, and protein measurements for assessing the state of unbalanced growth in the G_2-arrested, drug-treated subpopulation. In contrast, cells in the G_1 phase, that maintained normal levels of both RNA and protein, had survival values 40 times higher than the G_2 subpopulation. These results indicate that correlated analyses of DNA, RNA, and protein represent a very useful approach for studying mechanisms and control of the cell cycle.

8.1 Cell staining

Protocol 9. Fluorochrome labelling of DNA, RNA, and protein

1. Centrifuge ethanol-fixed cells prepared by *Protocol 1* and aspirate fixative.
2. Resuspend cells in staining solution containing HO 33342 (0.5 μg/ml), FITC (0.1 μg/ml) and pyronin Y (PY, 1.0 μg/ml) in PBS.[a] Adjust cell density to 7.5×10^5 cells/ml.
3. Allow stained cell sample to stand at room temperature for at least 1 h prior to analysis

[a] Stock solutions of Hoechst 33342 (HO 33342, Molecular Probes, Inc.) are prepared in distilled water at 1.0 mg/ml as stated above. Purified pyronin Y (Polysciences), also known as pyronin G or gelb, is a red fluorescent dye that binds preferentially to ribosomal RNA. Stock solutions of PY are prepared in dimethyl sulphoxide (DMSO) at 1.0 mg/ml and stored in a refrigerator for at least one year. DMSO may freeze at this temperature, but rapid thawing of the solution at 37°C prior to each use does not appear to produce any noticeable effect on PY–RNA stainability. Fluorescein isothiocyanate (FITC, isomer 1) (BBL, Microbiology Systems) is a green–yellow fluorescent dye that is often used as a fluorescent conjugate bound to antibodies. However, in this technique FITC is used in the free dye form as a general cellular protein stain. Only FITC of good quality and purity should be used for cell staining. FITC is prepared in 95% ethanol at 1.0 mg/ml by vortexing, and then this solution is diluted to about 10 μg/ml with distilled water. A fresh solution should be made just prior to use since FITC solutions do not appear to be stable for a long period of time.

8.2 Three-laser FCM setup for analysis of DNA–RNA–protein

The three-laser excitation flow system previously described (34) is used with the laser beams tuned to (333.6–363.8 nm), 457.9 nm, and 530.9 nm. The beams are focused on to the cell stream at three different points, with spacing of about 250 μm. The three-colour fluorescence detector consists of coloured glass and dichroic colour separating filters for measuring blue (400–500 nm), green (515–575 nm), and red (>580 nm) fluorescence emission from each stained cell. These emission measurement regions were preselected for each dye so that HO-bound DNA (blue), FITC-protein (green), and PY-bound RNA (red) fluorescence, respectively, are monitored at the UV, 457.9 nm and 530.9 nm excitation points on the cell stream. Fluorescence measurements are correlated on a cell-to-cell basis. Fluorescence signals and fluorescence ratios signals (20) are input to signal-processing electronics for subsequent storage (i.e., list mode fashion) in a DEC PDP-11/23 computer, and subsequently displayed as single-parameter frequency distribution histograms and two-parameter contour diagrams. Histograms for forward angle light scatter detected from 0.5 to 2.0°, three colours of fluorescence, and two fluorescence ratios are calculated from the data which were collected and stored in list mode for a given stained population of cells.

Gated analysis, as described previously (20), can be used to derive numerical data (i.e., relative values) for comparison of the ratios of RNA to DNA and RNA to protein for cells within specific regions of the cell cycle (37, 38). Using such analysis, a gate or window may be set between preselected regions (i.e., channel number x to channel number y) of the DNA histogram, encompassing, for example, cells in G_1 phase. Reprocessing of the data can then provide the numerical mean ratio value for cells within the G_1 region of the cell cycle. Since data are collected and stored in list mode fashion, data for any of the individual measurements can be retrieved in a similar manner for cells in any phase of the cell cycle. Potentially, gated analysis can also be performed on any of the distributions, i.e., protein or RNA content, and data can be correlated as described above.

8.3 Analysis of three-colour stained cells

The technique was used to characterize cycling CHO cells, and to correlate cellular DNA, RNA, and protein, as well as the ratios of RNA/DNA and RNA/protein throughout the cell cycle (see *Figure 4*). It is evident in *Figure 4A* and *B* that G_1 cells with a low RNA protein content do not enter the S-phase. Detailed characterization of the critical RNA and protein thresholds for entry of G_1 cells into S-phase were presented by Darzynkiewicz *et al.* (40). The RNA-to-protein contents (*Figure 4C*) are well correlated and consistent throughout the cell cycle.

Analysis of the cellular RNA/DNA ratio in relation to DNA content (*Figure 4E*) reveals a characteristic pattern reflecting changing rates of DNA replication and transcription during the cell cycle. Thus, during G_1 when DNA content is stable, cells accumulate increased quantities of RNA, but at different rates so that the G_1 phase is quite heterogeneous with respect to

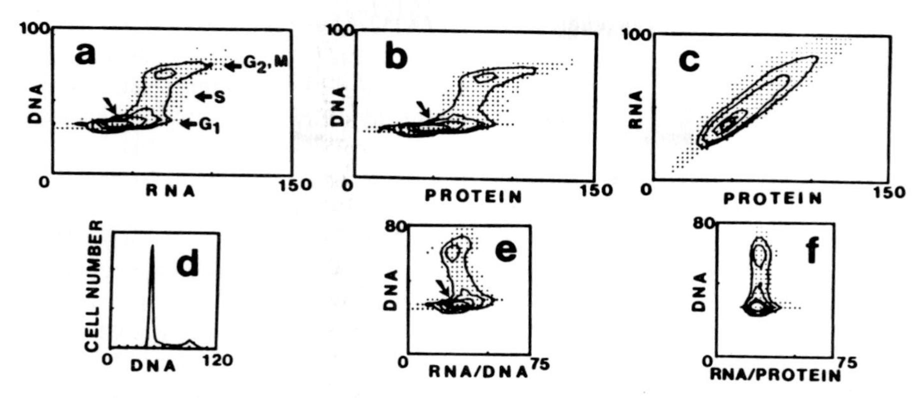

Figure 4. Bivariate contour distributions (a, b, c, e, and f) and a DNA frequency histogram (d) showing the distribution of exponentially growing CHO cells with respect to DNA, RNA, and protein content. The sequential contours represent increasing isometric levels equivalent to 5, 10, 50, 250, 500, and 1500 cells, respectively; 4×10^4 cells were measured per sample. The arrows indicate the threshold RNA or protein content of G_1 cells; cells with RNA or protein below the threshold values did not immediately enter the S phase (40).

RNA. However, during progression through S-phase, the rate of DNA replication exceeds RNA accumulation, giving rise to a non-vertical, negative slope of the S-phase cell cluster. Cells in G_2+M have RNA/DNA ratios in the same range as the majority of the G_1 cells. (For details see ref. 37.)

8.4 Controls for optimal colour resolution

As in any multicolour, fluorochrome-labelling technique, the dye concentrations should be maintained as low as possible to prevent dye–dye fluorescence interference, but sufficiently high enough to reflect quantitative labelling of the cellular content of interest. These criteria are determined empirically for each cell type. In this protocol the amount of the green-FITC and the red-PY fluorescence, as determined by the dye concentrations, must be adjusted accordingly to prevent red–green fluorescence interference, since the spectra for FITC and PY do overlap to some extent. In order to minimize potential blue (HO), green (FITC), and red (PY) fluorescence interference, a three laser, sequential excitation scheme (34) was devised. Also the 457.9 nm laser line, instead of the 488.0 nm line that is usually employed, was used for excitation of FITC in order to minimize excitation of PY. The PY was excited at 530.9 nm to minimize excitation of FITC. (For more details see also refs 37 and 38.)

In order to test the reliability of the protocol for quantitative determinations of cellular DNA, RNA, and protein it is usually necessary to analyse cells stained with only one of each of the dyes and compare results with those obtained for the same cells stained with all three dyes in combination. In all cases cells should be excited by all three laser beams and fluorescence analysis be performed in the blue, green, and red channels to determine the degree of non-specific fluorescence in each channel. In a previous study analysis of populations of CHO cells stained with all three dyes provided distribution profiles for DNA, protein, and RNA that were nearly identical to distributions obtained for cells stained with only one of each respective dye (37). Cells stained with all three dyes showed a slight decrease in (FITC) green fluorescence and a slight increase in (PY) red fluorescence as compared with respective single dye stained cells. This was due in part to energy transfer from FITC to PY. The DNA profile was similar, indicating no change in HO–DNA fluorescence intensity in the three-colour stained cells. Data were all obtained using the same electronic gain and laser power settings for the three lasers. Cells pretreated with RNase prior to PY staining and analysis showed an 85–90% decrease in red fluorescence compared with distributions for non-RNase-treated cultures.

9. Correlated analysis of cyclin B1 and cell cycle

Quantitation of proliferation markers is a rapidly growing area of study in cell biology. Such studies have been initiated to obtain information on those mechanisms that facilitate progression of cells through the various phases of the cell cycle. Of particular interest is the temporal expression of cyclin proteins, the regulatory subunits of cell cycle-dependent kinases. The B cyclins are important in regulating progression through mitosis. These proteins begin to accumulate during G_2, reach a maximum in M phase and are degraded during anaphase (41–43). Antibodies to cyclin B1 enable the immunofluorescence assay of expression of the cyclin during the cell cycle by flow cytometry.

9.1 Cell staining

The cell staining technique is similar to that described by Traganos *et al.* (44) with only minor changes.

Protocol 10. Cyclin B1 and DNA content labelling

1. Fix cells in 80% ethanol for at least 2 h.
2. Centrifuge cells at 200 *g* for 8 min.
3. Aspirate off 80% ethanol.
4. Add 2 ml of 0.5% Triton X 100 in filtered PBS for 5 min.
5. Add 2 ml of filtered phosphate buffered saline (PBS) and centrifuge cells at 200 *g* for 8 min.

6. Wash cells an additional time with 2 ml of (PBS) and centrifuge cells at 200 *g* for 8 min.
7. Incubate the cells with purified mouse anti-human cyclin B1 monoclonal antibody (mAb) (PharMingen) diluted 1:150 to equal 0.667 μg/ml antibody with 1% filtered bovine serum albumin (BSA; Calbiochem) in PBS. Add 100 μl/10^6 cells of the diluted B1 mAb to the sample and incubate at 4°C overnight.
8. Wash cells with PBS containing 1% BSA, centrifuge cells at 200 *g* for 8 min.
9. Incubate the cells with FITC-conjugated goat anti-mouse IgG antibody (Caltag) diluted 1:100 to equal 1 μg antibody with 1% BSA in PBS. Add 100 μl/10^6 cells of the diluted F-GAM to the sample and incubate for 30 min in the dark.
10. Add 2 ml of PBS and centrifuge cells at 200 *g* for 8 min.
11. Resuspend cells in 10 μg/ml PI and 50 μg/ml RNase. Incubate at room temperature for 30 min.

9.2 Analysis by flow cytometry

The cell sample is analysed as described for analysis of PI–FITC stained samples for DNA–protein content determinations above. Laser excitation is set at 488 nm and the red (DNA content) and green (cyclin B1 expression) is measured. *Figure 5* shows the results of analysis of asynchronous HeLa cells (A) and cells treated with nocodazole (B) to arrest cells in mitosis. The

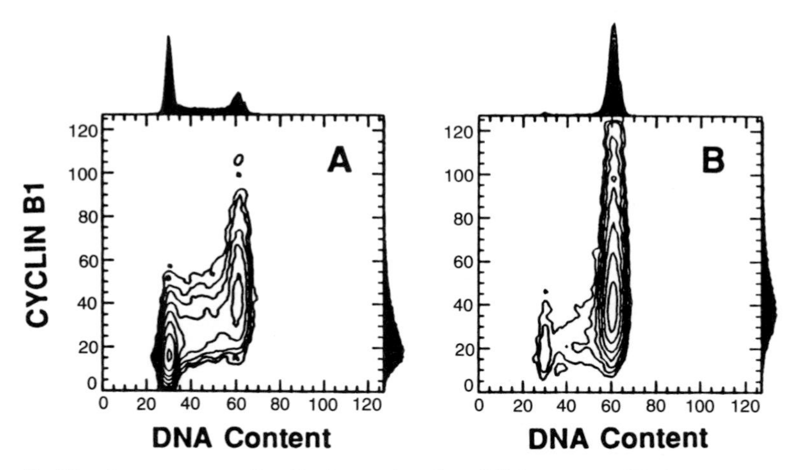

Figure 5. Bivariate contour distributions showing DNA content (PI fluorescence) and cyclin B1 expression (*y* axis) for asynchronous HeLa cells (A) or for cells treated for 16 h with nocadozole (B).

bivariate distributions in *Figure* 5A and B both show a significant increase in the expression of cyclin B1 in the G_2/M cells; however, expression is more clearly enhanced in the mitotic-arrested cells. Traganos *et al.* (44) have used this technique to examine cultures of MOLT-4 cells treated continuously with staurosporine under conditions in which the cells were undergoing polyploidization.

Acknowledgements

The author wishes to thank Leah Bustos, Judith Dickson, Shanna Minter, Robb Habbersett, and John Steinkamp for assistance in analysis of some of the data presented in this chapter. This work was supported by NIH Grant R24 RR06758, the Los Alamos National Flow Cytometry Resource funded by the Division of Research Resources of NIH (Grant P41–RR01315), and the Department of Energy.

References

1. Gurley, L. R., Enger, M. D., and Walters, R. A. (1973). *Biochemistry*, **12**, 237.
2. Hedley, D. W., Friedlander, M. L., and Taylor, I. W. (1985). *Cytometry*, **6**, 327.
3. Hedley, D. W. (1990). In *Flow Cytometry* (ed. Z. Darzynkiewicz and H. A. Crissman), pp. 139–47. Academic Press, San Diego, CA.
4. Latt, S. A. (1973). *Proc. Natl. Acad. Sci. USA*, **70**, 3395.
5. Arndt-Jovin, D. J. and Jovin, T. M. (1977). *J. Histochem. Cytochem.*, **25**, 585.
6. Crissman, H. A., Stevenson, A. P., Kissane, R. J., and Tobey, R. A. (1979). In *Flow cytometry and sorting* (ed. M. R. Melamed, P. F. Mullaney, and M. L. Mendelsohn), pp. 243–61. Wiley, New York.
7. Krishan, A. (1990). In *Flow Cytometry* (ed. Z. Darzynkiewicz and H. A. Crissman), pp. 121–5. Academic Press, San Diego, CA.
8. Vindelov, L. and Christensen, I. J. (1990). In *Flow Cytometry* (ed Z. Darzynkiewicz and H. A. Crissman), pp. 127–37. Academic Press, San Diego, CA.
9. Zelenin, A. V., Poletaev, A. I., Stephanova, N. G., Barsky, V. E., Kolesnikov, V. A., Nikitin, S. M., Zhuze, A. L., and Gnutchev, N. V. (1984). *Cytometry*, **5**, 348.
10. Gallop, P. M., Paz, M., Henson, S. A., and Latt, S. A. (1984). *Biotechniques*, **1**, 32.
11. Darzynkiewicz, Z. (1990). In *Flow Cytometry* (ed. Z. Darzynkiewicz and H. A. Crissman), pp. 285–98. Academic Press, San Diego, CA.
12. Hirons, G. T., Fawcett, J. J., and Crissman, H. A. (1994). *Cytometry*, **15**, 129.
13. Fried, J., Doblin, J., Takamoto, S., Perez, A., Hansen, H., and Clarkson, B. (1982). *Cytometry*, **3**, 42.
14. Crissman, H. A., Hofland, M. H., Stevenson, A. P., Wilder, M. E., and Tobey, R. A. (1988). *Exp. Cell. Res.*, **174**, 388.
15. Krishan, A. (1987). *Cytometry*, **8**, 642.
16. Alabaster, O., Tannenbaum, E., Habbersett, M. C., Magrath, I., and Herman, C. (1978). *Cancer Res.*, **38**, 1031.

17. Krishan, A., Ganapathi, R., and Israel, M. (1978). *Cancer Res.*, **38**, 3656.
18. Vindelov, L. L., Christensen, I. J., Jensen, G., and Nissen, N. I. (1983). *Cytometry* **3**, 328.
19. Crissman, H. A., Gadbois, D. M., Tobey, R. A., and Bradbury, E. M. (1991). *Proc. Natl. Acad. Sci. USA*, **88**, 7580.
20. Crissman, H. A. and Steinkamp, J. A. (1973). *J. Cell Biol.*, **59**, 766.
21. Stohr, M., Vogt-Schaden, M., Knobloch, M., Vogel, R., and Futterman, G. (1978). *Stain Technol.*, **53**, 205.
22. Pollack, A. (1990). In *Flow Cytometry* (ed Z. Darzynkiewicz and H. A. Crissman), pp. 315–23. Academic Press, San Diego, CA.
23. Roti-Roti, J. L., Higashikubo, R., Blair, O. C., and Vygur, H. (1982). *Cytometry*, **3**, 91.
24. Crissman, H. A. and Steinkamp, J. A. (1982). *Cytometry*, **3**, 84.
25. Steinkamp, J. A. and Crissman, H. A. (1974). *J. Histochem., Cytochem.*, **22**, 616.
26. Dolbeare, F., Gratzner, H., Pallavacini, M., and Gray, J. W. (1983). *Proc. Natl. Acad. Sci. USA*, **80**, 5573.
27. Gratzner, H. (1982). *Science*, **218**, 474.
28. Bohmer, R. M. (1979). *Cell Tissue Kinet.*, **12**, 101.
29. Bohmer, R. M. and Ellwart, J. (1981). *Cytometry*, **2**, 31.
30. Ellwart, J. and Dohmer, P. (1985). *Cytometry*, **6**, 513.
31. Kubbies, M. and Rabinovich, P. S. (1983). *Cytometry*, **3**, 276.
32. Poot, M., Hoehn, H., Kubbies, M., Grossman, A., Chen, Y., and Rabinovitch, P. S. (1990). In *Flow Cytometry* (ed. Z. Darzynkiewicz and H. A. Crissman), pp. 185–206. Academic Press, San Diego, CA.
33. Crissman, H. A. and Steinkamp, J. A. (1987). *Exp. Cell Res.*, **173**, 256.
34. Steinkamp, J. A., Stewart, C. C., and Crissman, H. A. (1982). *Cytometry*, **2**, 226.
35. Steinkamp, J. A. and Stewart, C. C. (1986). *Cytometry*, **7**, 566.
36. Habbersett, R., Crissman, H. A., and Jett, J. H. (1990). *Cytometry* (Suppl. 4), Abst 528B, p. 88.
37. Crissman, H. A., Darzynkiewicz, Z., Tobey, R. A., and Steinkamp, J. A. (1985). *Science*, **228**, 1321.
38. Crissman, H. A., Darzynkiewicz, Z., Tobey, R. A., and Steinkamp, J. A. (1985). *J. Cell Biol.*, **101**, 141.
39. Crissman, H. A., Wilder, M. E., and Tobey, R. A. (1988). *Cancer Res.*, **48**, 5742.
40. Darzynkiewicz, Z., Traganos, F., and Melamed, M. R. (1980). *Cytometry*, **1**, 98.
41. Gallant, P. and Nigg, E. A. (1992). *J. Cell Biol.*, **117**, 213.
42. Murray, A. W., Soloman, M. J., and Kirschner, M. W. (1989). *Nature*, **349**, 281.
43. Pines, J. and Hunter, T. (1989). *Cell*, **58**, 833.
44. Traganos, F., Gong, J., Ardelt, B., and Darzynkiewicz, Z. (1994). *J. Cell Physiol.*, **158**, 535.

3

Analysis of flow cytometric DNA histograms

PETER S. RABINOVITCH

1. Introduction

Analysis of the DNA content of cells and their distribution within G_1, S and G_2/M phases of the cell cycle is one of the earliest applications of flow cytometry. Because of its simplicity, DNA content analysis remains a commonly used and powerful technique. Chapters 2 and 8 within this volume describe additional variations and enhancements of the original flow cytometric techniques; however, mathematical analysis of single parameter DNA content histograms remains a common element in many of these methods. Recent interest in the clinical application of DNA ploidy and cell cycle measurements from DNA histograms has stimulated improvements in the methods and models used to derive these measurements, and guidelines for the implementation of DNA cytometry have been established to help ensure greater accuracy in implementation of these techniques. This chapter describes a protocol consistent with these guidelines for DNA content histogram analysis and discusses some of the principles in methods of analysis. References in Sections 3 and 4 to the DNA Cytometry Consensus Conference (1) are denoted by an asterisk.

2. Histogram analysis

Protocol 1. DNA histogram analysis

Materials

- Cells stained with a DNA-specific dye (see Chapters 2 and 8)
- A flow cytometer operated according to manufacturers' recommendations
- Cell cycle analysis software and a compatible computer (usually a personal computer) for analysis. The software may be that supplied by the cytometer manufacturer, or software available from independent vendors.

Protocol 1. *Continued*

A. *Data acquisition*

1. Tune the cytometer, and assure performance (e.g., Coefficient of Variation (CV = peak standard deviation/mean $\times$ 100) and linearity) by using appropriate standards or controls.
2. Obtain DNA fluorescence histogram(s) by analysis of the stained cells or nuclei. Critical aspects are:
   (a) Cell Number: a minimum of 10 000 events in a DNA content cell cycle analysis is recommended, although detection of DNA aneuploid populations may be possible from histograms with fewer cells or nuclei.
   (b) CV: a low CV is important for accurate resolution of DNA aneuploid populations, as well as for accurate S and G_2/M phase measurements. CVs should generally be below 8.
   (c) G_1 peak position and number of histogram channels: the lowest G_1 population should be accumulated in channel numbers greater than 30, and probably above 50, in order that there be a sufficient number of channels for adequate peak resolution and cell cycle analysis. To the left of the G_1 peak, sufficient debris should be collected in the histogram to allow software programs sufficient data to construct a model for debris compensation; setting a lower limit of data acquisition at a channel that corresponds to DNA index (DI) 0.1 is recommended. The DI = DNA value/ G_1 DNA value. The right portion of the histogram is also important in order to assess the extent of aggregation, and, if software aggregation modelling is applied, these data are essential to the proper use of such modelling. The DNA Cytometry Consensus Guidelines recommend collecting data up to DI 6.0, or even to DI 10.0 if an aneuploid population with DI>2.0 is present (1).
   (d) Histogram linearity: departures from linearity can produce non-standard G_2/G_1 ratios, altered DNA indices, and difficulty in computer modelling of aggregation. Instrument linearity should be determined on a regular basis, using standard particles or cells, and appropriate corrections made (1).
   (e) DNA content standards: variations in DNA dye binding between a cell type used for an external DNA content standard (such as lymphocytes or nucleated red blood cells) and the cells in a DNA content histogram can result in ambiguity in the correct diploid DNA content, and thus, small differences in staining intensity relative to 'standards' cannot be interpreted as evidence of DNA aneuploidy. In analysis of tissue samples, the best DNA content standard is the normal tissue component that represents the nor-

mal counterpart of the neoplastic cells. In the case of formalin fixed tissue, variability in fixation and DNA–dye accessibility prohibits any reproducibility in the position of the standard peak, and it is recommended that the left-most peak from paraffin-embedded material be assumed to represent the DNA diploid population (1).

B. *Cell cycle model fitting*

1. Confirm the initial software estimates or manually assign the initial estimates needed for cell cycle analysis, particularly the G_1 and G_2 peak means and CV. If more than one ploidy population is present (in cells from malignant tissue, for example), initial analysis should be performed with a cell cycle model that has the same number of cell cycles (G_1, S and G_2/M phase components) as the estimated number of ploidy populations. Remember that the formal definition of DNA aneuploidy requires that at least two distinct peaks are present (1, 2). If a population of apoptotic cells is present, these can usually be fit as an additional peak.
2. Select a model shape for fitting the S-phase distribution. In general, choose the simplest shape that will suffice (see below). In the case of cell populations that have been experimentally cell cycle arrested or synchronized, a more flexible shape for the S phase model, such as Fox's method (3), may be utilized.
3. Include background debris in the cell cycle model. Histogram dependent (not simple exponential) debris modelling should be used. Use aggregation modelling, unless aggregates have been sucessfully excluded by pulse shape discrimination (peak/area or width/area gating).
4. Fit the chosen cell cycle model to the data.

C. *Histogram interpretation and/or re-analysis*

1. Examine the results of fitting with the chosen cell cycle model, and reassess the validity of the model and the results. Evaluate the validity of the chosen number of cycling populations. DNA aneuploid populations should generally be considered to be present when such cells constitute >5% of cells (or nuclei), after correction for aggregates and debris. If necessary, refit abnormally elevated G_2/M fractions as separate tetraploid populations.
2. Further disaggregate any samples in which the analysis indicates the presence of an excessive proportion of aggregated cells, and reanalyse the sample.
3. Revise any features of the cell cycle model, if necessary, and refit the data. Revisions may include reduction of the complexity of the model (e.g. simpler shape to the S-phase distribution, G_1 and G_2/M peak CVs constrained to be equal, G_2/G_1 ratio constrained to a value near 2.0).

Protocol 1. *Continued*

This is especially useful when the histogram is complex or of poor quality, or when overlapping peaks give rise to aberrant fitting results.

4. Evaluate the accuracy of the S and G_2/M phase measurements. Criteria include:
   (a) The percentage of aneuploid cells in an aneuploid S phase is optimally above 20%.
   (b) The percentage background aggregates and debris (BAD) is optimally below 20%.
   (c) The CV should be less than 8%.
   (d) Skewing of the G_1 peak(s) should not overlap the S phase(s).
   (e) Evaluate the chi square of the fit, or preferably, more sophisticated measures of the statistical confidence in the results obtained with the fitting model.
   (f) For additional indication of reliability, compare results obtained by fitting different variations in the cell cycle model. This not only allows the determination of the best fitting model, but a greater range of results obtained with different model variations indicates less reliability in the fitting results.

3. Cell cycle modelling

3.1 The DNA content cell cycle

Quiescent and non-replicating cells have a G_0 or G_1 DNA content. Cell replication begins with DNA synthesis, or S phase, during which the DNA content gradually increases from the G_1 DNA content to twice this value, the G_2 and mitotic cell (G_2/M) DNA. Completion of mitosis returns the DNA content to the G_1 value. Thus, in a growing cell population, the distribution of DNA content measured by flow cytometry consists of cells with G_1 and G_2/M DNA content and a broad distribution of DNA contents between G_1 and G_2/M. In the theoretical distribution, these S phase cells are easily identified between G_1 and G_2/M DNA contents (*Figure 1A*, heavy solid lines). In actual histograms, however, the uncertainty in measurements and Gaussian broadening of G_1, S, and G_2/M phase distributions results in considerable overlap between G_1 cells and early S phase cells and G_2/M cells and late S phase cells (*Figure 1A*, curves). When cells with an abnormal DNA

Figure 1. The difference between a histogram from a 'perfect' flow cytometer with no errors in measurement (heavy solid lines) and the Gaussian broadening of the G_1, S, and G_2/M phase components that is encountered in all real analyses (light solid lines). Actual data points are displayed as dots, broadened curves are fit with the Dean and Jett polynomial S phase model (ref. 4). The dashed line shows the overall fit of the model to

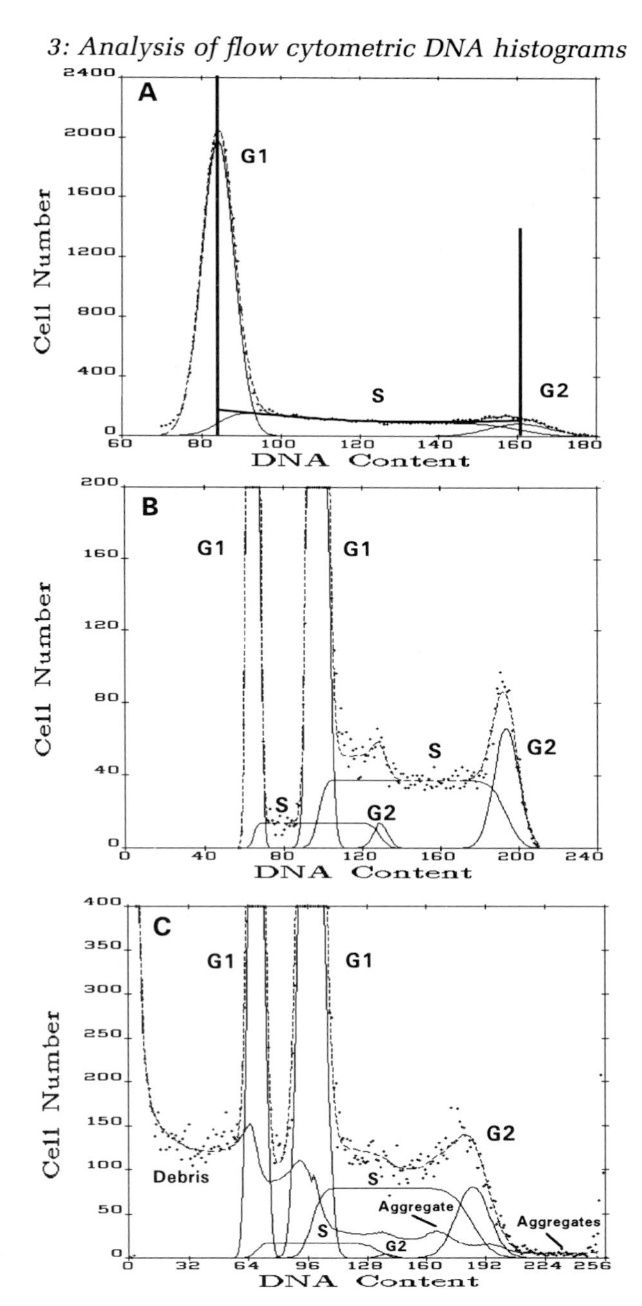

the data. B illustrates the same model fitting, but to a histogram that has overlapping diploid and aneuploid cell cycles. C shows a histogram with diploid and aneuploid cells, but with the addition of debris resulting from extraction of nuclei from paraffin, together with aggregates of cells with cells and cells with debris. A solid line shows the combined distribution of background aggregates and debris.

content are present in tissue, second G_1, S, and G_2/M phases are present, since the aneuploid cells are almost always accompanied by a component of cells with normal, diploid DNA content (for example stromal fibroblasts, capillary endothelial cells and lymphocytes). The overlap between the diploid and aneuploid cell cycles can be variable, but adds to the complexity of the histogram analysis (*Figure 1B*). Cell cycle analysis is applied to DNA histograms in order to extract measurements of DNA content and G_1, S, and G_2)M phase fraction.

3.2 Model components

DNA content histograms require mathematical analysis in order to extract the underlying G_1, S, and G_2/M phase distributions; methods for this analysis have been developed and refined over the past two decades. The most flexible and accurate methods of cell cycle analysis are based upon building a mathematical model of the DNA content distribution, and then fitting this model to the data using curve-fitting methods. The best established model, proposed by Dean and Jett (4) is based on the prediction that the cell cycle histogram is a result of the Gaussian-broadening of the theoretically perfect distribution (*Figure 1A*). The underlying distribution can be reconstructed or 'deconvoluted' by fitting the G_1 and G_2/M peaks as Gaussian curves and the S-phase distribution as a Gaussian-broadened distribution. As originally proposed, the shape of this broadened S phase distribution is modelled as a smooth second-order polynomial curve (a portion of a parabola). The model can be simplified by using a first-order polynomial curve (a broadened trapezoid) or a zero-order curve (a broadened rectangle). Especially complex S phase shapes, for example those resulting from synchronized populations of cells in S phase, can be modelled by an extension of the above methods that allows the S phase distribution to be the combination of a second order polynomial and a Gaussian curve (3). Curve-fitting models are usually fit to the histogram data by use of least squares analysis. The fitting model is used to generate a mathematical expression, or function, for the predicted histogram distribution. The variables in the function (usually more than a dozen) are adjusted to give the optimum concordance (lowest least-squares difference) between the fitting model and the observed data. Curve-fitting methods are also the least dependent on the initial or 'starting parameters' used to begin the fitting process.

The models used by the least squares fitting method can be directly extended to analysis of two (*Figure 1B*) or even three overlapping cell cycles. The overlapping model components are mathematically deconvoluted to yield individual cell cycle estimates. Debris and aggregates, when present, can be fit by appropriate models in order to compensate for their effects on the cell cycle analysis (*Figure 1C*). If a population of apoptotic cells is present, these can usually be fit as an additional peak.

3.3 Debris modelling

The purpose of debris modelling is to compensate for the effects of debris that may underlie the cell cycle distribution. Such debris may be a result of damaged or fragmented cells or nuclei, or nuclei cut during sectioning of tissue from a paraffin block. The earliest debris-fitting models consisted of an exponentially declining curve (i.e., $a \cdot e^{-kx}$), however, this simple model has been found to be generally unsatisfactory. In particular, fragments of debris are always smaller than the DNA content of an intact nucleus and thus debris extends only leftward of the G_1, S, or G_2 from which it is derived. In modelling this debris, the shape of the debris curve is thus dependent upon where the peaks in the DNA histogram are; models of this kind are termed 'histogram dependent'. It is recommended that

* for mathematical modelling, simple exponential debris models are not reliable, and so histogram-dependent debris models should be used. Sophisticated mathematical modelling techniques currently available with DNA content analysis software allow elimination of debris from otherwise good quality histogram data; their use is strongly recommended, particularly for S-phase estimations.

Figure 2A illustrates modelling of a simple exponential curve to the debris region left of the G_1 peak. This curve predicts too much debris over the S and G_2/M phases and results in a zero % S-phase fraction (SPF) estimate. Histogram-dependent exponential debris modelling is shown in *Figure 2B*. The background debris curve drops rapidly from the left side of the G_1 peak to the right side of the G_1 peak because most of the debris is produced from G cells and extends only leftward of the G_1 DNA content. In this case, the fitting model yields an 6.8% SPF. The histogram-dependent exponential model does not, however, provide an ideal fit to debris from paraffin-extracted tissue; this can be seen in *Figure 2B*, and especially in *Figure 2D*. Nuclei cut during sectioning of a paraffin block yield a characteristic pattern of debris from the cutting of nuclei that are in the path of the knife. This results in a random distribution of fragment sizes, ranging from small to almost full size. There tends to be a slightly greater fraction of small and large fragments produced during cutting (resulting from an excess of relatively small and large produced by cutting the rounded ends of the nuclei), and the distribution of sliced fragments exhibits a concave rather than a flat distribution (5). This same shape may also be seen in fresh specimens that are minced with a scalpel, forced through mesh, or otherwise cut. A mixture of histogram-dependent exponential debris and cut nucleus debris can be analysed by combining these two models. *Figure 2C* (S phase 8.2%) and *E* illustrate the close fit of this combined model to debris in histograms derived from nuclei extracted from paraffin-embedded tissue. This model provides good correlation with known S-phase values in experimental data (6), and improved

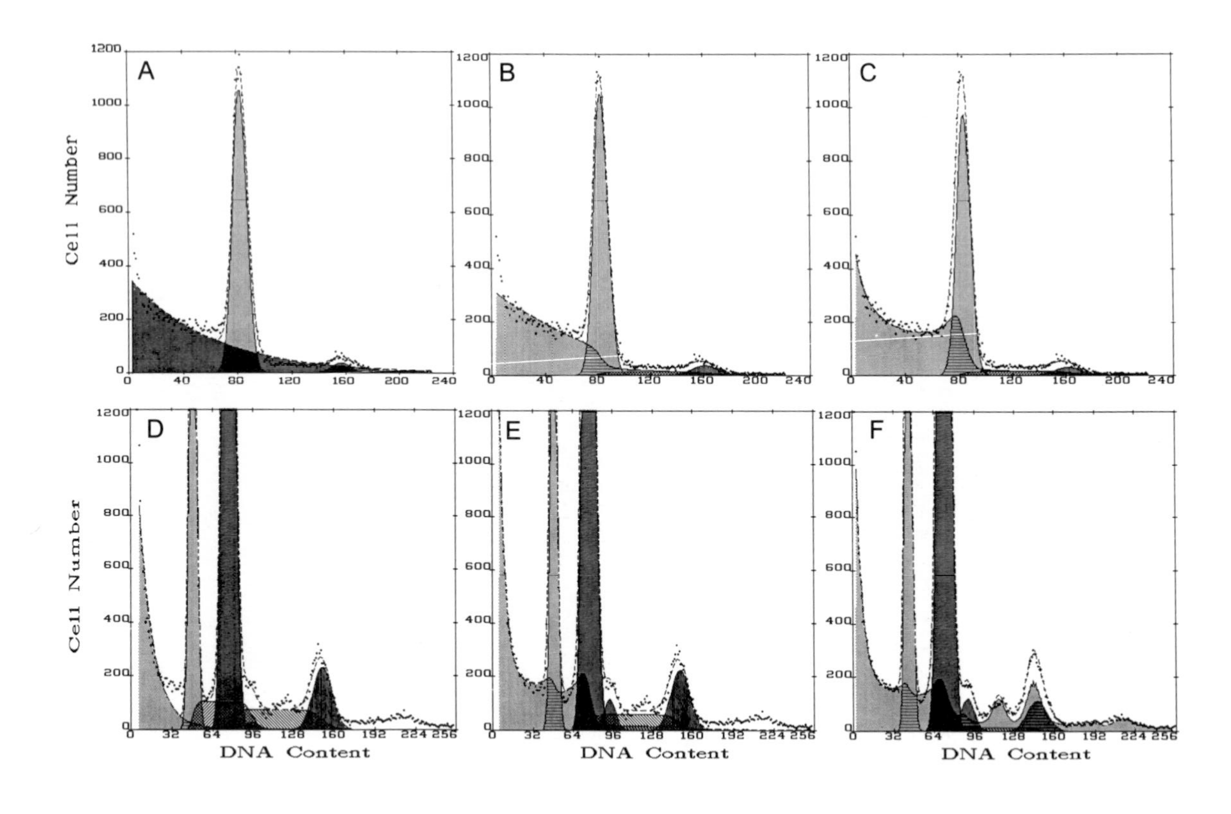
A
B
C
D
E
F
Cell Number
Cell Number
DNA Content
DNA Content
DNA Content

Figure 2. Simple exponential debris model (A), histogram-dependent exponential debris model (B) and sliced nucleus debris modelling (combined exponential and cut nucleus histogram dependent models) (C) applied to a DNA diploid histogram from a paraffin-preserved tissue sample containing degenerating cells. Debris curves are seen at the left of the histogram, continuing to various extents rightward. The S phase fractions (SPF) resulting from the cell cycle analysis are 0, 6.8% and 8.2%, respectively. D–E show analysis of paraffin-preserved tissue with DNA aneuploidy, fit with histogram-dependent exponential debris model (D), sliced nucleus debris modelling (combined exponential and cut nucleus histogram dependent models) (E) and sliced nucleus modelling with aggregation modelling (F). Note the overestimation of diploid SPF in (D) (/// cross-hatching), and the overestimation of aneuploid SPF (\\\ cross-hatching) in both (D) and (E) due to the presence of aggregation. The aggregates, are well fit in (F), including the diploid G_1–aneuploid G_1 doublet near channel 110, and the higher-order aggregates to the right of the aneuploid G_2/M. Cell cycle analyses in this and other figures were performed using the MultiCycle program written by the author (Phoenix Flow Systems, San Diego, CA).

---

prognostic strength of SPF measurements in clinical studies (7, 8). The combined model is also relatively insensitive to the left and right endpoints chosen for the fitting region, and this reduces interoperator variation in cell cycle analysis results (7).

3.4 Aggregation modelling

Careful inspection of many histograms will reveal evidence of cell aggregation: although an aggregate of two G_1 cells (a doublet) will have the same DNA content as a G_2/M cell, and may be overlooked, diploid triplets will be seen at DI 3.0, quadruplets at DI 4.0, etc. In addition to these more obvious aggregates, S and G_2/M cells and debris also can aggregate with G_1 cells and with each other. Aneuploid cells, if present, may aggregate with each other and with diploid cells and debris. The overall effect on the histogram can be complex. The DNA Cytometry Consensus Guidelines caution that

* the presence of aggregates can affect detection of DNA aneuploid peaks and may cause major errors in S-phase determination.

In the past, the primary approach to detection of aggregates has been to distinguish the altered pulse shape that they may produce when analysed by a flow cytometer using a focused laser beam. Although this method may be successful when examining spherical and relatively uniform cells, with cells or nuclei that are heterogeneous in shape or oblong, as is common in many epithelial malignancies, the pulse-shape gating may have variable success (6). For example, if a G_1 cell doublet passes through the flow cell parallel to the laser beam, rather than perpendicular to it, then the fluorescence pulse-shape cannot be distinguished from that of a G_2/M cell. A G_1 doublet may not be distinguishable from an oblong G_2/M cell. Triplets or quadruplets may form a 'spheroid' without a longer axis, and thus may not be distinguishable from a single large cell. Aggregated nuclei have no intervening cytoplasm to keep the nuclei from being tightly opposed, and the pulse shape of aggregated nuclei is less discernible from that of whole cell aggregates.

The pulse-shape analysis is most commonly performed by plotting fluorescence peak (or pulse width) vs. fluorescence area from each cell. A diagonal line is then drawn, making the assumption that aggregates will fall below the line, having a lower pulse peak for a given pulse area than non-aggregates. Placement of the diagonal line is subject to user interpretation. For epithelial cells, the G_1 and G_2/M peak/area or width/area relationship may be variable, and assessing the position of the diagonal line may be more difficult.

As an extension of the 'histogram dependent' modelling approach, a computer model can be applied that allows a generalized approach to the fitting of aggregation in DNA histograms (6, 9, 10). The basis of this model is the simple assumption that any two particles, i.e., elements of the histogram, have a certain probability of aggregating with each other. The net aggregate distribution has a shape that is formed from the composite of all possible aggregate combinations. The aggregate distribution can be added to the cell cycle and debris models and the combined model is fit to the observed data by using least-squares fitting. Examples of histograms fit with aggregation modelling are shown in *Figures 2F* and *3*. Hardware gating may affect the aggregate distribution in the histogram and may alter the 'expected' aggregate relationships. Therefore, pending future detailed study of these interactions,

* use of software aggregation modelling should be restricted to data collected without hardware gating.

As for debris modelling, this also means avoiding the use of light scatter gates.

3.5 Complexity of the model

The cell cycle model used for histogram analysis is composed of one G_1, S, and G_2 phase per cycling DNA content, debris, and when needed aggregation modelling. Within this framework, there may be greater or less complexity in

- S phase shape
- G_2 CV: allowed to be independently evaluated, or constrained to be the same as the G_1 CV
- G_2/G_1 ratio: allowed to vary, or constrained to the usual value obtained in the laboratory

---

Figure 3. Application of aggregation modelling to a DNA aneuploid histogram from a carcinoma of the breast. Panel A shows the cell cycle analysis without aggregation modelling. A peak is present near channel 65 that might be mistaken for a second DNA aneuploid population. B (10× *y* axis scale) shows the same histogram analysed with aggregation modelling added to the sliced nucleus debris (horizontal shading). Diploid and aneuploid S phases are shown by diagonal hatching. The total fit is indicated by the short-dashed line. C (20× *y* axis scale) shows the individual components of the back-

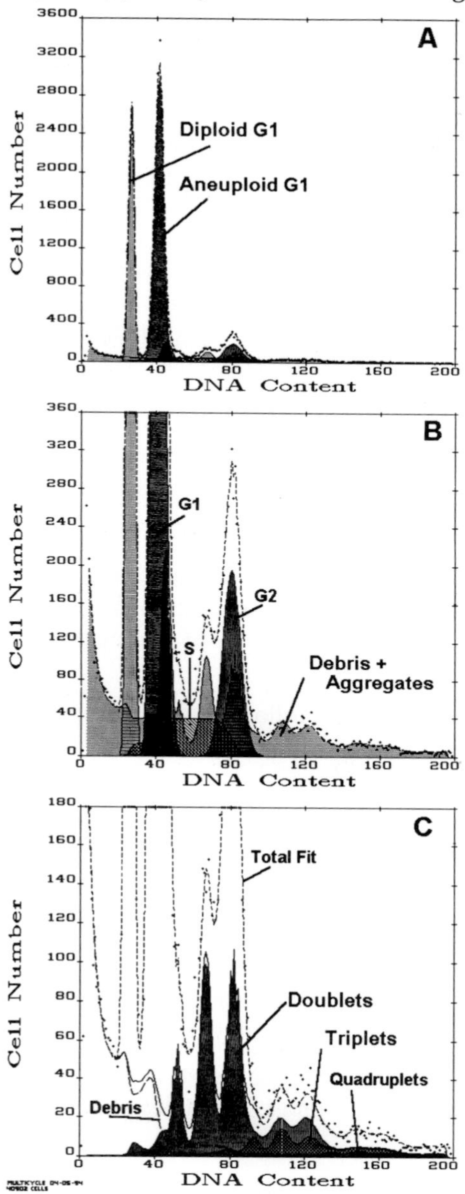

ground fit. At the left of the histogram is the sliced nucleus debris (long-dashed line). The debris plus aggregates is shown as a solid line. The doublet distribution (vertical stripes) is complex in shape, reflecting the fact that all histogram components (diploid G_1, S, and G_2/M, aneuploid G_1, S, and G_2/M) aggregate with each other. The triplet distribution (checkered) has higher DNA content overall than the doublets, but there is extensive overlap. The quadruplet distribution (solid) is even higher in DNA content, but overlaps the triplet distribution to a large extent.

* CV of aneuploid G_1 and G_2 phases (if present): different from the diploid G_1 and G_2 CV, or fit with one CV that best fits both diploid and aneuploid peaks.

When the quality of a DNA histogram is optimal, then the fitting model may be used with its fullest complexity, as this confers the greatest freedom for the model to match the data most exactly. Difficulties may arise, however, as the DNA histogram becomes less perfect (lower numbers of events, non-Gaussian and non-symmetrical peak shapes, increasing proportions of debris), more complex (greater number of ploidy populations, increased aggregation) or more ambiguous (overlapping peaks, overlapping aggregates and peaks). In such cases, allowing greater freedom and complexity in the model can be counterproductive. Artefacts may be fit as cell cycle components, and results may be less, rather than more, accurate. In these circumstances, it is often advisable to introduce constraints in the model's complexity.

4. Reliability of histogram analysis results

* Not all histograms are adequate for identification of DNA aneuploid populations, or for the estimation of S phase—when the histogram is inadequate, due to high CVs, debris, or aggregates, it should be reported as inadequate.

Factors to consider include the following.

4.1 CV and peak shape

Broader CVs impair the detection of near-diploid DNA aneuploid populations and increase the overlap of G_1 and G_2 populations with the S-phase distribution, making analysis more difficult.

* In general, the CV of normal diploid cells in a histogram should be less than 8%.

The presence of a skewed peak shape also has an adverse impact on S-phase measurement, as the skew may overlap the S-phase distribution.

* When a skewed G_1 peak or a peak with a 'tail' on the right side extends visibly into the S phase, S phase estimates should be used with extreme caution.

4.2 Quantitation of aggregates and debris

The proportion of events analysed by the flow cytometer that consist of cell or nuclear debris or aggregates is an indication of the degree to which these events complicate the assessment of DNA content and proliferative fractions in the histogram. To address the need for a quantitative measure of aggregates and debris, the DNA Cytometry Consensus Conference created a parameter termed Background Aggregates and Debris (BAD), defined as the proportion of the histogram events between the leftmost G_1 and the rightmost G_2/M that is modelled as debris or aggregates. The DNA Consensus Conference Guidelines recommend that

* if a sample has a high percentage of aggregates (>10% as determined by manual counting), it should be further disaggregated by mechanical means or rejected for analysis. The percentage aggregates and debris should be evaluated for each histogram . . . it is recommended that >20% histogram background aggregates and debris are unsatisfactory for S phase analysis.

4.3 Proportion of aneuploid cells

When aneuploid cells are present, accurate S phase estimation requires that they should be greater than 15–20% of the total cells. This is because overlap of cells from a large diploid population into the S phase region of the aneuploid population can impair the accuracy of estimation of the aneuploid S phase; furthermore, even if the two populations do not extensively overlap, debris and aggregates from the more abundant population still overlap the cell cycle of the rarer population.

4.4 Statistical measures derived from the fitting model

The most conventional parameter of 'goodness of fit' of the least square cell cycle analysis is the chi square statistic, χ^2. The χ^2 statistic may thus be useful to indicate the extent to which the fitting model matches the histogram distribution, lower values of the χ^2 being better. However, the χ^2 statistic is also affected by the number of cells acquired in the histogram and by the relative proportion of cell cycle vs. background distribution represented in the histogram. It is thus unlikely that any absolute criteria can be formulated based upon χ^2.

The non-linear least squares method of fitting can provide a statistical indicator of the confidence of each cell cycle parameter. The non-linear least squares technique optimizes the fit by adjusting each fitting parameter to minimize the value of χ^2; the rate of change in χ^2 as the parameter is adjusted is an indicator of the uncertainty in the parameter, estimated using an 'error matrix' (11). Error estimates based on this method may be useful to indicate a range of confidence in cell cycle measurements (6, 12).

4.5 Sensitivity of S and G_2/M phase measurements to variations in the fitting model

The extent to which the cell cycle parameters vary depending upon the model chosen ('inter-model' variation) is an indicator of the range of uncertainty in the estimates. This indicator is more reflective of potential errors due to non-optimal choice of the fitting model than are the estimates described above.

References

1. Shankey, T. V., Rabinovitch, P. S., Bagwell, C. B. *et al.* (1993). *Cytometry*, **14**, 472.
2. Hiddeman, W., Schumann, J., Andreeff, M., Barlogie, B., Herman, C. J., Leif, R. C., Mayall, B. H., Murphy, R. F., and Sandberg, A. A. (1984). *Cytometry*, **5**, 445.
3. Fox, M. H. (1980). *Cytometry*, **1**, 71.
4. Dean, P. and Jett, J. (1974). *J. Cell Biol.*, **60**, 523.
5. Bagwell, C. B., Mayo, S. W., Whetstone, S. D., *et al.* (1991). *Cytometry*, **12**, 107.
6. Rabinovitch, P. S. (1993). In *Clinical flow cytometry principles and applications* (ed. K. D. Bauer, R. E. Duque, and T. V. Shankey), pp. 117–42. Williams and Wilkins, Baltimore, MD.
7. Kallioniemi, O.-P., Visakorpi, T., Holli, K., Heikkinen, A., Isola, J., and Koivula, T. (1991). *Cytometry*, **12**, 413.
8. Kallioniemi, O.-P., Visakorpi, T., Holli, K., Isola, J. J., and Rabinovitch, P. S. (1994). Automated peak detection and cell cycle analysis of flow cytometric histograms. *Cytometry*, **16**, 250.
9. Rabinovitch, P. S. (1990). *Cytometry Supp.*, **4**, 27.
10. Bagwell, C. B. (1993). In *Clinical flow cytometry principles and applications* (ed. K. D. Bauer, R. E. Duque, and T. V. Shankey), pp. 41–62. Williams and Wilkins, Baltimore, MD.
11. Bevington, P. R. (1969). *Data reduction and error analysis for the physical sciences*, pp. 153–60. McGraw-Hill, New York.
12. Rabinovitch, P. S. (1993). *Multicycle software, advanced version*, Phoenix Flow Systems, San Diego.

4

Determination of the cell proliferative fraction in human tumours

ROSELLA SILVESTRINI, MARIA GRAZIA DAIDONE, and AURORA COSTA

1. Introduction

During the last two decades, increasing effort has been made to obtain as much information as possible on the cell proliferative fraction of individual clinical tumours in order to increase the knowledge on tumour biology and to provide clinicians with data on consecutive series of patients. As a consequence, an impressive number of reports has supported the relevance of cell kinetics as a tool to investigate growth pattern, biological heterogeneity and clinical progression of human tumours (1–4).

To make the determination of the cell proliferative fraction of human tumours feasible on consecutive series of patients, investigators have addressed their attention to specific aspects of the complex phenomenon of proliferation and tumour growth. In particular, kinetic characterization has focused on the proliferating cell compartment, that is, on cells which transit through the four phases of the cell cycle, which are generally responsible for tumour growth, and which are most susceptible to the action of therapeutic agents.

Biologists and pathologists have used several approaches to determine the cell proliferative fraction. Such approaches are based on different rationales and employ different methods of evaluation, such as morphometric, immunocytochemical, cytometric or autoradiographic detection, according to the professional background of the investigators (5). They are intended to analyse and quantify either the whole proliferating fraction or discrete fractions of cells in specific cell cycle phases, mainly in the S phase.

2. Methods for procurement and preparation of tumour specimens

2.1 Procurement of tumour tissue

Preparation of tumour specimens to be processed for cell kinetic studies varies according to the tumour type, the site of localization, and the proliferative index to be determined. However, two prerequisites must be satisfied. The first deals with sampling of tissue specimens from solid tumours, which should be made by pathologists. Second, the time between removal of the tissue and initiation of specific methodological procedures should be kept to a minimum, and the tissue must be kept moist meanwhile.

The latter prerequisite is favoured when the cell kinetic laboratory is close to surgical rooms and to the pathology department. Otherwise, for the determination of proliferation indices requiring viable cells for the active incorporation of nucleic acid precursors, tissue samples may be placed in sterile, disposable dishes or vials containing a phosphate-buffered saline solution (PBS), Hank's balanced salt solution (HBSS), or culture medium (we use RPMI 1640 or MEM-alpha). Samples must be chilled in ice when the interval between excision and incubation with nucleic acid precursors is long. In any case, incubation must be carried out within 2–3 h. A longer time markedly decreases the basic metabolic activity of the excised tissue and compromises the reliability of results.

Cell aspirates (effusions or bone marrow) must be collected in tubes or syringes containing 2‰ heparin without preservatives, and the suspension should be mixed well. Fine-needle aspirates should be placed in tubes containing PBS.

2.2 Preparation of specimens from solid tumours

All the following statements apply mainly to cell kinetic determinations which involve the use of viable cells incorporating nucleic acid precursors.

Differences in tissue architecture among the different histotypes, the importance of cell-to-cell interactions, and the danger of selection of cell clones not representative of the overall cell subpopulations of the original tumour tissue make the proposition of a 'standard' procedure for dispersing the tissue somewhat unreliable and also questions the use of cell suspensions from solid tumours. Comparison of the different procedures used to obtain cell dispersion indicates the importance of using small fragments in order to obtain reliable proliferation indices. The only exception is solid localizations of systemic tumours, such as non-Hodgkin's lymphomas, for which cell suspensions obtained from involved lymph nodes are adequately representative of the original tumour cell population.

2.2.1 Mincing

Tumour specimens, free of fat and necrotic or haemorrhagic areas, are minced in sterile dishes (100 mm diameter) containing medium or PBS into fragments of 1–3 mm^3. The use of scissors should be avoided due to compression injury to the cells adjacent to the cut surface. Instead, a very sharp scalpel blade should be used. Care should be taken to make clean cuts through the tissue and to avoid tearing the material. Once the tissue is minced, the fragments can be transferred for incubation with nucleic acid precursors into vials previously prepared with complete culture medium.

2.2.2 Dispersion of tissue to obtain cell suspensions

Tumour tissue is minced in sterile dishes (100 mm diameter) containing medium or PBS by passing two very sharp cataract knives smoothly in opposite directions through the tissue. Once the tissue is dispersed, the fragments are processed differently for mechanical disaggregation or enzymatic digestion.

i. Mechanical disaggregation

Tumour fragments are transferred to a test tube containing a magnetic stirring bar and culture medium. This procedure, which is advisable for involved lymph nodes from non-Hodgkin's lymphomas, relies on the dispersion of cells from the tissue fragments by the shearing forces of circularly flowing medium during gentle magnetic stirring. A different procedure to obtain single cell suspensions, which is mainly efficient for malignant melanomas, requires that tumour fragments are transferred to a plastic bag containing culture medium and treated for a minute or less with a Stomacher Lab-Blender 80 (Seward Laboratory). Cells released after the two procedures are collected and filtered through 100-mesh gauze for removal of cell clumps. Cell suspensions are then centrifuged at 1000–1200 *g* for 20 min and resuspended in culture medium. An aliquot is counted in a haemocytometer chamber (see Chapter 1) to adjust the appropriate cell number for the different cell kinetic determinations.

ii. Enzymatic disaggregation

Since there is no standard procedure for tissue dispersion, there is no standard mixture for enzymatic disaggregation. An often used enzymatic mixture is that proposed by Slocum *et al.* (6). Briefly, tumour fragments are incubated in RPMI 1640 with 10% fetal calf serum, 0.8% collagenase II, 0.002% DNase I at 37°C for 2 h in a shaking water bath. Another procedure, frequently used for lung tumours, requires the use of a mixture of 0.25% collagenase and 0.01% hyaluronidase at 37°C for 1 h (7). At the end of the incubation, the released cells are collected, washed, filtered and processed as described above for mechanical dispersal.

2.3 Treatment of cell aspirates

2.3.1 Cell aspirates from solid lumps, washings and effusions

This category of samples includes:

- fine-needle aspirates collected in PBS

- bladder washings collected in PBS
- pleural and ascitic fluids collected in heparinized solutions

Samples are centrifuged at 700–1000 *g* and washed in fresh PBS. Cell suspensions are then centrifuged, resuspended in culture medium and counted as described in Section 2.2.2(i).

2.3.2 Cell aspirates from bone marrow

Bone marrow aspirates are collected in preservative-free heparin, resuspended in ice-cold culture medium (RPMI 1640) and processed for Ficoll Hypaque density separation. Specimens are layered onto Ficoll Hypaque (specific gravity, 1.077) (Pharmacia LBK) and centrifuged at 1000–1200 *g* for 20–30 min. The light-density cells are recovered, washed twice in RPMI 1640, centrifuged and resuspended in fresh RPMI 1640 and enumerated as described in Section 2.2.2(i).

3. Proliferation indices

Several experimental approaches (complementary and/or alternative) have been proposed for cell kinetic determinations on large and consecutive series of human tumours (5). They can be grouped according to the rationale on which they are based and according to the aspect of the cell cycle to be analysed: discrete fractions of cells in specific cell cycle phases, mainly in the S phase, or the entire proliferating cell fraction.

3.1 Incorporation of nucleic acid precursors

This approach uses labelled pyrimidine bases, such as [^{3}H]thymidine or [^{14}C]thymidine (8), or halogenated analogues, such as bromo- or iododeoxyuridine (9, 10), which are specifically incorporated into DNA during S phase. Such measurements involve autoradiographic or immunocytochemical techniques and can be performed on histological sections or cytospins obtained from surgical or biopsy specimens. **Quantification of S-phase cells is expressed as the percentage of DNA precursor-incorporating cells over the total number of tumour cells.** The main advantages of these approaches are that they can be carried out readily on a large number of cases and, as *in situ* procedures, that it is possible to discriminate tumour from non-tumour cells. The limitation to widespread use is the **requirement for fresh tumour material.** This constraint has been partially overcome by the availability of kits for [^{3}H]thymidine (distributed by Ribbon) and for bromodeoxyuridine (distributed by Amersham), which, besides guaranteeing the methodological standardization of the first steps, facilitate performance in peripheral institutions.

3.1.1 [^{3}H]Thymidine labelling index

For determination of the labelling index with [^{3}H]thymidine (the most frequently used radionuclide for cell kinetic studies), tumour material should

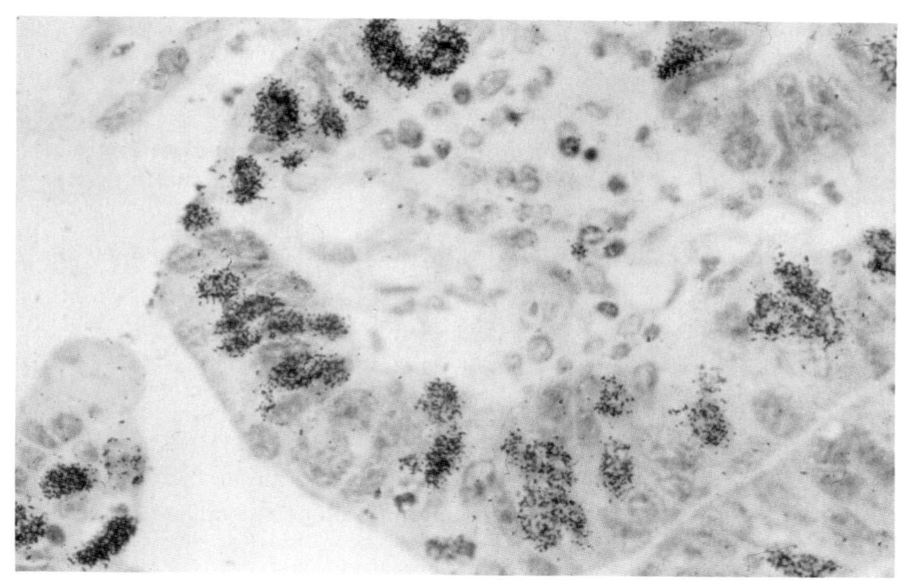

Figure 1. Microphotograph of *in vitro* [^{3}H]thymidine labelled cells using the autoradiographic technique on sections from a colon adenocarcinoma.

be **processed within 60 min of surgery or biopsy.** The procedures are described in *Protocols 1* and *2* for tumour fragments (*Figure 1*) and cell suspensions, respectively.

Protocol 1. Determination of [^{3}H]thymidine labelling index on tumour fragments

Reagents

- MEM-alpha (Whittaker) with 20% fetal calf serum (Gibco) and antibiotics
- [^{3}H]Thymidine (Amersham): 222 kBq/ml; specific radioactivity 925 GBq/mmol
- 2′-Deoxythymidine (Merck): 0.25 mg/ml; M = 242.23 g/mol
- Gelatin (Difco Laboratories): 5‰
- Emulsion K5 (Ilford)
- D-19B Developer: dissolve 2 g photo-rex (Merck), 72 g anhydrous sodium sulphite (Carlo Erba), 9 g hydroquinone (Merck), and 4 g potassium bromide (Merck) in a final volume of 1 litre of distilled water. Make fresh every week, store in a brown bottle at 4°C. Always filter before use
- Fixer: dissolve 240 g sodium thiosulphate (Carlo Erba) in 150 ml of distilled water. Dissolve 15 g anhydrous sodium sulphite (Carlo Erba), 8 g boric acid (Merck), and 15 g aluminium potassium sulphate (Carlo Erba) in 13 ml glacial acetic acid (Carlo Erba) and distilled water. Mix the two solutions and add distilled water to a final volume of 1 litre. Make fresh every week, store in a brown bottle at 4°C. Always filter before use
- Mayer's haematoxylin: dissolve 1 g haematoxylin indicator (Merck), 49 g aluminium potassium sulphate (Carlo Erba), 0.2 g sodium iodate (Merck), 49 g chloral hydrate (Carlo Erba), and 1 g citric acid (Merck) in a final volume of 1 litre of distilled water. It can be used only 15 days after preparation and will last 2–3 months
- Eosine Y (Merck): 0.25%

Protocol 1. *Continued*

A. *Incubation with the nucleic acid precursor*

1. Randomly pick up 8–10 tumour fragments (see Section 2.2.1).
2. Put the fragments in 1.9 ml of complete culture medium and add 0.1 ml of [^{3}H]thymidine. Incubate at 37°C for 1 h in a shaking water bath.
3. Remove the culture medium and wash fragments to remove the unincorporated precursor with ice-cold PBS.
4. Fix the fragments in 10% buffered formalin for 2–24 h.
5. Leave overnight in 80% ethanol.

B. *Histological procedure*

1. Dehydrate the samples and embed them in paraffin.
2. Cut 4 μm-thick sections and mount these on gelatin-covered slides.

C. *Autoradiography*

1. Deparaffinate the slides as follows:
   (a) xylene: 1 h, repeat twice
   (b) absolute ethanol: 5 min, repeat twice
   (c) 95% ethanol: 5 min
   (d) 80% ethanol: 30 min
   (e) tap water: 2 min
2. Rinse in distilled water: 15 min, repeat four times.
3. Wash with a solution containing unlabelled thymidine for 40 min.
4. Rinse in tap water (2 min), then in distilled water (15 min, repeat four times), and leave the slides in distilled water (pH 7) until they are dipped in the photographic emulsion.

Perform steps 5–11 in the darkroom at 23°C, 60% humidity, with a Safelamp fitted with a Kodak Safelight Filter (no.1, red) and a 15 pearl lamp.

5. Cover the wet slides with emulsion diluted 1:1 with distilled water (pH 7) and maintained at a constant temperature in a 40°C water bath. Remove the emulsion on the opposite side of slides and place them in histological plates.
6. Let the emulsion dry for 10 min at 4°C and then for 2 h at room temperature.
7. Place the slides in a light-tight box with anhydrous $CaSO_4$, or more

simply wrap the histological plates in black paper and enclose the package in an open-ended polyethylene bag.

8. Store in a refrigerator at 4°C for 3 days.
9. Develop the slides in D-19B developer at 19°C (5 min).
10. Fix in fixer solution at 19°C (10 min).
11. Rinse the slides in tap water (5 min, repeat twice), and then rinse in distilled water (5 min).

D. *Staining*

1. Stain with Mayer's haematoxylin (30 min).
2. Wash briefly in distilled water and then rinse in tap water (10 min).
3. Stain with eosine (3 min). Then wash briefly in distilled water (repeat twice).

Protocol 2. Determination of the [^{3}H]thymidine labelling index in cell suspensions

Reagents

- Bouin's solution: 600 ml picric acid solution (Merck), 200 ml 40% w/v formaldehyde (Carlo Erba), and 40 ml glacial acetic acid (Carlo Erba)
- MEM-alpha or RPMI 1640 medium plus 20% fetal calf serum and antibiotics
- [^{3}H]Thymidine: 75 kBq/ml, specific radioactivity 185 GBq/mmol

A. *Incubation with the nucleic acid precursor*

1. Prepare a cell suspension containing at least 10^6 viable cells (see Section 2.2.2) in 1.9 ml of complete culture medium.
2. Add 0.1 ml of [^{3}H]thymidine. Incubate at 37°C for 1 h in a shaking water bath.
3. Remove the culture medium by centrifugation (1000–1200 *g* for 7 min at 4°C), wash cells to remove unincorporated precursor with cold PBS, and cytocentrifuge (100 000 cells per slide, centrifuge for 3 min at 30 *g*).
4. Fix the cytospins in Bouin's solution for 1 h.
5. Wash briefly with 80% ethanol (repeat three times).
6. Leave overnight in 80% ethanol.

B. *Autoradiography*

1. Hydrate slides with distilled water (15 min, repeat four times).

Protocol 2. *Continued*

2. Rinse in tap water (2 min).

Steps 3–7 are the same as C3–7 described for tumour fragments in *Protocol 1*.

8. Store in a refrigerator at 4°C for 2 days.

Steps 9–11 are the same as C9–11 described for tumour fragments in *Protocol 1*.

C. *Staining*

The procedure is the same as described in *Protocol 1*.

3.1.2 Bromo- and iododeoxyuridine labelling index

For the determination of the labelling index with halogenated analogues of pyrimidine bases, *in vitro* incubation of tumour tissue (9) or *in vivo* injection in patients is employed (10). Tumour material should be **processed within 60 min of surgery or biopsy**. *Protocol 3* describes the approach to determine the bromodeoxyuridine labelling index for tumour fragments (*Figure 2*), and *Protocol 4* covers the methodology for cell suspensions.

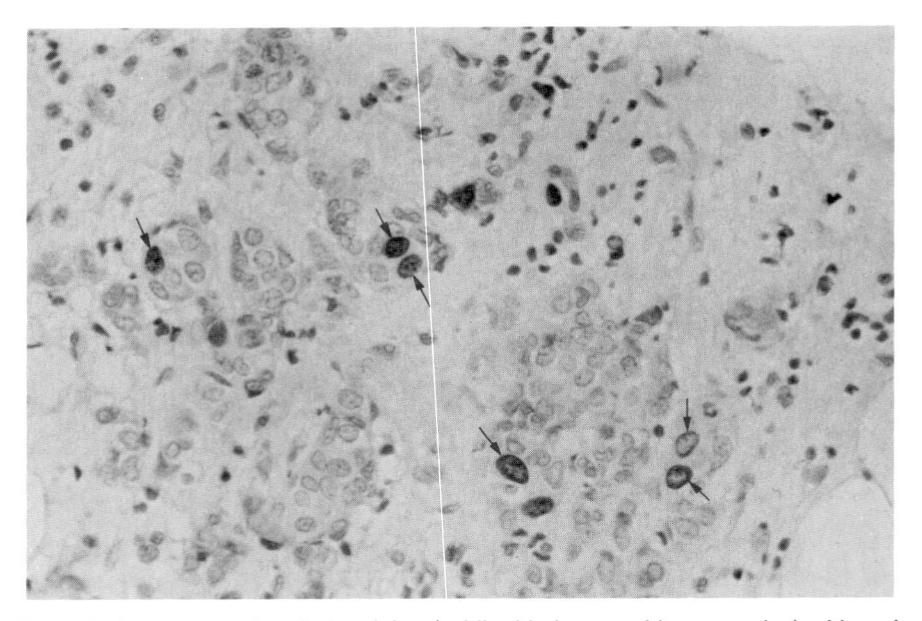

Figure 2. Immunocytochemical staining (avidin–biotin–peroxidase complex) with anti-bromodeoxyuridine antibody to detect DNA synthesis in a sample of breast adenocarcinoma (arrows).

Protocol 3. Determination of the bromodeoxyuridine labelling index in tumour fragments

Reagents

- Bromodeoxyuridine (Sigma): 1.5 mg in 20 ml PBS. Prepare a solution 2.44 μM and store aliquots of 1 ml at −20°C. Warm to 37°C before use
- RPMI 1640 with 20% fetal calf serum and antibiotics
- mAb anti-bromodeoxyuridine (Becton Dickinson): dilution 1:20 in PBS + 0.05% Tween 20
- Anti-mouse IgG biotinylated antibody (Vector Laboratories): dilution 1:200 in PBS + 0.03% bovine serum albumin (BSA)
- Avidin–biotin complex Vectastain kit
- Diaminobenzidine (Sigma): 6 mg + H_2O_2 (36 vol.) 0.01 ml in 10 ml of PBS

A. *Incubation with the halogenated analogue*

1. Randomly pick up 8–10 tumour fragments (see Section 2.2.1).
2. Put fragments in 1.9 ml of complete culture medium and add 0.1 ml of stock solution of bromodeoxyuridine. Incubate at 37°C for 30–60 min in a shaking water bath. Avoid culture media containing thymidine, such as MEM-α, F-10, F-12.
3. Remove the culture medium and wash in PBS for 15 min at 37°C.
4. Process tissue for frozen sectioning or paraffin embedding as required. For a better morphological preservation, we suggest using paraffin-embedded material. For such a procedure, fix fragments in 10% buffered formalin for no longer than 12 h.
5. Leave overnight in 80% ethanol.

B. *Histological procedure*

The procedure is the same as described in *Protocol 1.*

C. *Immunocytochemistry*

1. Deparaffinate slides as follows
   (a) xylene: 15 min
   (b) absolute ethanol: 10 min
   (c) 95% ethanol: 5 min
   (d) 80% ethanol: 5 min
   (e) distilled water: 2 min
2. Block endogenous peroxidase with 36 vol. H_2O_2 (3% in PBS for 10 min at room temperature in the dark), then rinse briefly in PBS.
3. Denaturate DNA (at room temperature) with:
   (a) 3 M HCl: 20 min
   (b) 0.1 M borax buffer, pH 8.5: 10 min

Protocol 3. *Continued*

(c) PBS: 5 min, repeat twice
(d) pronase E, 0.5 mg/ml: 5 min
(e) PBS: 5 min, repeat twice.

4. Block non-specific sites by incubation in PBS containing 0.5% Tween 20 and 0.3% BSA for 20 min at room temperature.

Perform the following steps (5–10) in a humid chamber at room temperature.

5. Incubate with mAb against the halogenated analogue (1 h).
6. Wash in PBS for 5 min (repeat twice).
7. Incubate with secondary mAb for 30 min.
8. Wash in PBS for 5 min (repeat twice).
9. Incubate with avidin–biotin complex 1:50 in PBS for 30 min.
10. Wash in PBS for 5 min (repeat twice).

D. *Staining*

1. Stain with a diaminobenzidine solution for 5 min in the dark.
2. Wash in tap water for 2 min.
3. Counterstain with Mayer's haematoxylin for 2 min.
4. Wash briefly in distilled water (repeat twice).
5. Rinse in tap water for 5 min.
6. Dehydrate slides, mount with a permanent mountant and coverglass.

Protocol 4. Determination of the bromodeoxyuridine labelling index in cell suspensions

Reagents

See *Protocol 3*.

A. *Incubation with the halogenated analogue*

1. Prepare a cell suspension containing at least 10^6 viable cells (see Section 2.2.2) and incubate in the freshly prepared labelling medium previously described (see *Protocol 3* for tumour fragments) at 37°C for 30–60 min in a shaking water bath.
2. Remove the culture medium and wash in PBS for 15 min at 37°C and cytocentrifuge (100 000 cells per slide, centrifuge for 3 min at 30 *g*).

3. Fix the cytospins in ice-cold 70% ethanol for 30 min.
4. Store at 4°C until use (from 30 min up to 1 week).

B. *Immunocytochemistry*

1. Hydrate slides with PBS for 10 min.
2. Block endogenous peroxidase as described for tumour fragments in *Protocol 3*.
3. Denaturate DNA (room temperature) with:
   (a) 3 M HCl: 20 min
   (b) 0.1 M borax buffer, pH 8.5: 10 min
   (c) PBS: 5 min (repeat twice)

Steps 4–10 are the same as described for tumour fragments in *Protocol 3*.

C. *Staining*

The procedure is the same as described in *Protocol 3*.

3.2 Other measurements of cell proliferation

These approaches, which in some instances represent the natural evolution and the integration of morphometric and functional determinations, should provide information on the overall fraction of proliferating cells, that is, the growth fraction of the tumour. However, particularly for solid tumours, the sensitivity, specificity and reproducibility of these approaches are still being assessed. In fact, available information on their clinical relevance, although interesting, is still controversial and indicates the necessity for methodological verification and standardization through quality control assessments. Such measurements involve immunocytochemical techniques and are generally performed on histological sections obtained from surgical or biopsy specimens. **Quantification of the fraction of proliferating cells is expressed as the percentage of labelled cells over the total number of tumour cells.**

3.2.1 Ki-67 or MIB-1 index

Ki-67 is an antibody, first identified in phytohaemagglutinin-stimulated lymphocytes, that recognizes an antigen specifically expressed by proliferating cells (11). The antigen has a defined chemical structure (a dimeric, non-histone protein with a molecular weight of 356-395 kDa). Its function is still undefined but is not similar to that of DNA polymerase or topoisomerase II. Its presence has been detected in acetone-fixed frozen sections, and, more recently (12), in paraffin-embedded sections using the monoclonal antibody MIB-1. Determination of Ki-67 index is described in *Protocol 5*, and the MIB-1 index (*Figure 3*) is described in *Protocol 6*.

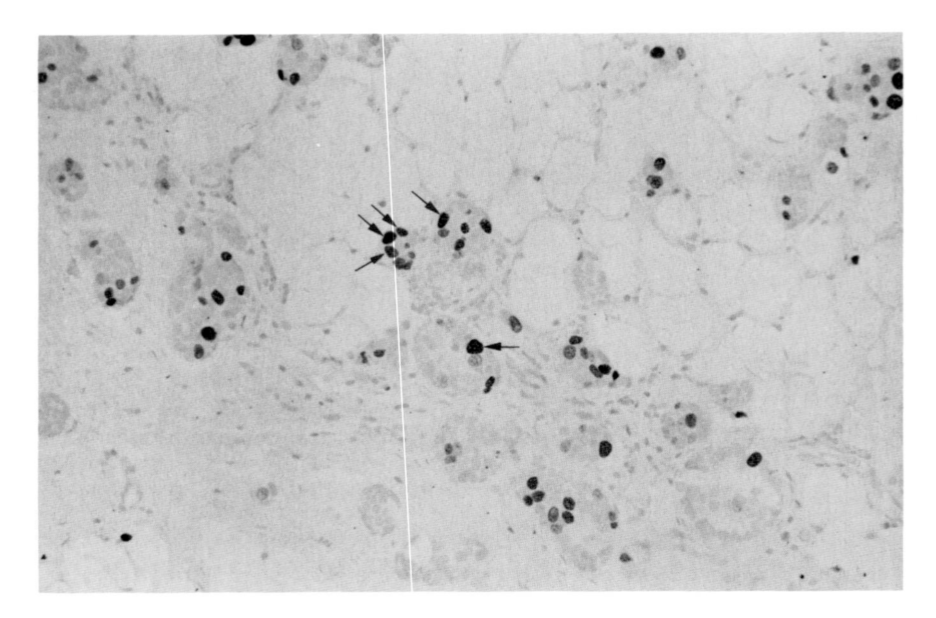

Figure 3. MIB-1 immunoreactivity in breast adenocarcinoma. Positive nuclei are visualized by an avidin–biotin–peroxidase complex (brown stain; see arrows) and counterstained with Mayer's haematoxylin.

Protocol 5. Determination of Ki-67 index on acetone-fixed frozen sections

Reagents

- mAb Ki67 (DAKO SpA): dilution 1:10 in PBS + 0.03% BSA
- Anti-mouse IgG biotinylated antibody (Vector Laboratories) dilution 1:200 in PBS + 0.03% BSA
- Avidin–biotin complex Vectastain Kit: dilution 1:50 in PBS

A. *Histological procedure*

1. Freeze tumour specimen.
2. Cut 5 μm-thick sections.
3. Fix in ice-cold acetone for 10 min.

B. *Immunocytochemistry*

1. Wash specimen in PBS.
2. Block endogenous peroxidase with 36 vol. H_2O_2 (3% in PBS) for 10 min at room temperature in the dark, then rinse briefly in PBS.

3. Block non-specific sites by incubation in PBS containing 3% BSA for 20 min at room temperature.

Perform steps 4–9 in a humidified chamber at room temperature.

4. Incubate with the mouse mAb Ki67 for 1 h.
5. Wash in PBS for 5 min (repeat twice).
6. Incubate with secondary mAb for 30 min.
7. Wash in PBS for 5 min (repeat twice).
8. Incubate with avidin–biotin complex for 30 min.
9. Wash in PBS for 5 min (repeat twice).

C. *Staining*

The procedure is the same as described in *Protocol 3.*

Protocol 6. Determination of MIB-1 index on paraffin-embedded sections

Reagents

- Poly-L-lysine (Sigma), 0.01%
- Citrate buffer, 10 mM, pH 6
- mAb MIB-1 (Immunotech S.A.): dilution 1:50 in PBS + 0.03% BSA
- Anti-mouse IgG biotinylated antibody (Vector Laboratories): dilution 1:200 in PBS + 0.03% BSA
- Avidin–biotin complex Vectastain kit: dilution 1:50 in PBS

A. *Histological procedure*

1. Fix fragments in 10% buffered formalin.
2. Dehydrate samples and embed them in paraffin.
3. Cut 3–5 μm-thick sections and mount on poly-L-lysine-covered slides.

B. *Immunocytochemistry*

1. Deparaffinate slides.
2. Block endogenous peroxidase as described in *Protocol 3.*
3. Incubate in a microwave oven in citrate buffer: boil for 5 min (repeat twice); cool for 30 min at room temperature, and wash briefly in PBS.
4. Block non-specific sites by incubation in PBS containing 3% BSA for 20 min at room temperature.

Perform steps 5–10 in a humid chamber at room temperature.

5. Incubate with mAb MIB-1 for 1 h.
6. Wash in PBS for 5 min (repeat twice).

Protocol 6. *Continued*

7. Incubate with secondary mAb for 30 min.
8. Wash in PBS for 5 min (repeat twice).
9. Incubate with avidin–biotin complex for 30 min.
10. Wash in PBS for 5 min (repeat twice).

C. *Staining*

The procedure is the same as described in *Protocol 3*.

3.2.2 Cyclin-proliferating cell nuclear antigen (PCNA) index

Cyclin–PCNA is a nuclear protein (with a molecular weight of 36 kDa) that is expressed specifically by all the proliferating cells. However, the type of fixation markedly affects the possibility to detect the expression of the antigen in cells from the different cell cycle phases (13). Expression of the protein is identified by an immunoperoxidase technique with the use of monoclonal antibodies (19A2 or PC10) on formalin-fixed, paraffin-embedded sections. *Protocol 7* provides the methodology for obtaining a cyclin–PCNA index.

Protocol 7. Determination of a cyclin–PCNA index

Reagents

- mAb 19A2 (Biogenex): dilution 1:20
- mAb PC10 (DAKO): dilution 1:100
- Anti-mouse IgM (Vector Laboratories) at a 1:200 dilution for 19A2
- Anti-mouse IgG2A (Amersham) at a 1:40 dilution for PC10
- Avidin–biotin Vectastain kit: dilution 1:50 in PBS

A. *Histological procedure*

The procedure is the same as described in *Protocol 1*.

B. *Immunocytochemistry*

1. Deparaffinate slides by the procedure described in *Protocol 3*.
2. Denature DNA at room temperature with:
   (a) 2 M HCl (for 20 min)
   (b) 0.1 M Borax buffer, pH 8.5 for 10 min
   (c) PBS for 5 min (repeat twice)
3. Block endogenous peroxidase with 36 vol. H_2O_2 (3% in PBS) for 10 min at room temperature in the dark, then rinse briefly in PBS.
4. Block non-specific sites by incubation in PBS containing 3% BSA for 20 min at room temperature.

Perform steps 5–10 in a humidified chamber at room temperature.

5. Incubate with primary antibodies in PBS + 0.03% BSA for 1 h.
6. Wash in PBS for 5 min (repeat twice).
7. Incubate with the secondary biotinylated antibody in PBS + 0.03% BSA for 30 min.
8. Wash in PBS for 5 min (repeat twice).
9. Incubate with avidin–biotin complex (for 30 min).
10. Wash in PBS for 5 min (repeat twice).

C. *Staining*

The procedure is the same as described in *Protocol 3*.

3.3 Methodological controls

The increasing use of determinations of the proliferative fraction in human tumours for clinical applications requires an assessment of the different methodological steps to provide investigators with standardized means to obtain reliable and consistent information on proliferation of individual tumours.

For solid tumours, preliminary studies have defined the critical points in sample preparation. Small fragments obtained from biopsy or surgical specimens represent the optimal sampling for incorporation studies. Conversely, at least for some tumour types such as breast cancer, core biopsies can underestimate tumour heterogeneity, and fine-needle biopsies are markedly dependent on the expertise of the operator, since the procedure must be minimally traumatic to avoid cell damage.

In addition, for the most widely used approaches, such as determination of the fraction of cells in the S phase by [^{3}H]thymidine incorporation or by flow cytometry (see Chapters 2 and 4), methodologies have been or will be standardized using quality control programmes (14, 15). Such programmes, which are ongoing in Italy, involve 15–20 independent centres including national cancer institutes and universities. The aim is to examine the reproducibility of the determination of the proliferation index in different laboratories through a periodic blind assessment. The final objective of these quality control programmes is to provide clinicians with a network of qualified and experienced laboratories.

3.4 Relationship between the different proliferation indices

Biological and clinical information provided by the different proliferation indices is not always superimposable, particularly when dealing with solid tumours. Consequently, for clinical applications, it is still not possible to define the most feasible, reproducible and least-expensive indicator which is able to guarantee accuracy and reliability in predicting biological and clinical aggressiveness. At present, the most widely adopted proliferation indices [^{3}H]thymidine labelling index, flow cytometric S-phase cell fraction and Ki-67 index—are based on different rationales and are characterized by different methodological complexities. This latter aspect could lead to the selection of a particular approach through opportunistic rather than carefully weighed clinical considerations. This could unreasonably preclude the use of kinetic variables, which require more complex and sophisticated manipulations, including the use of specific fixatives or incubation of viable tumour material with metabolic precursors.

The use of phenotypic proliferation markers (Ki-67 or MIB-1 and cyclin–PCNA index) or of markers based on phenomenological detection (flow cytometric S-phase cell fraction, see Chapters 2 and 4) may be most appropriate for clinical purposes, to acquire information in a relatively short time, and possibly when studying archival specimens (frozen or paraffin-embedded). However, most of these techniques still require a methodological standardization and assessment of intra- and interlaboratory reproducibility. Moreover, reagent specificity for the cellular target has been only partially defined for the easily assessable immunohistochemical procedures, and the sensitivity of such methods is sometimes inadequate, particularly for slowly proliferating tumours. Conversely, for complex procedures based on DNA precursor incorporation, the specificity in detecting S-phase cells is well known. As a consequence, without definitive information on target specificity or methodological sensitivity, a validation of the different approaches can be obtained only through appropriate studies on substantial case series of patients with the different neoplasms.

4. Predictive value of cell proliferative fraction measurements

Measurement of the proliferative fraction of tumour cell populations has become increasingly important as a complement to the clinicopathological findings in making treatment decisions. In fact, the clinical aggressiveness of tumours has consistently proved to be directly related to certain cell kinetic variables. Evidence is also emerging in favour of a relation between cell proliferation and the response to specific treatments in several human tumour types.

In general for any biological variable, and in particular for cell kinetics, its relevance as an indicator of prognosis or of treatment response should be analysed on different case series, that is, on patients homogeneous for pathological stage submitted possibly to only local–regional treatment for the former analysis, or to therapies homogeneous for type and intensity of drugs for the latter analysis. These separate analyses are made necessary by the confounding evidence that a rapid proliferation is indicative of opposite behaviours. In fact, a rapid proliferation correlates with a poor prognosis and a high biological aggressiveness but is also indicative of susceptibility to intensive cytotoxic treatment. Therefore, it could even be indicative of a greater chance of response to chemotherapy, although intrinsic drug sensitivity of the cells remains a major limitation in clinical response. Since the interpretation of clinical outcome as a function of proliferative rate is complex and could be affected by the two opposite trends observed for rapidly proliferating tumours, correlative analyses should be performed on case series treated differently, according to the specific objective of the study.

4.1 Prognosis

The marked independence of tumour cell proliferation rate from important prognostic factors, such as clinical or pathological stage, initially raised concerns about its relevance as a prognostic indicator. However, findings from retrospective correlative studies consistently showed that the proliferative fraction (mainly determined as autoradiographic or flow cytometric S-phase cell fraction) is an important indicator of clinical outcome in several tumour types at early stages, such as breast, head and neck, colorectal, gastric and non-small cell lung cancers and malignant melanomas (16). In addition, the association between a rapid cell proliferation and a poor prognosis is maintained independently of other traditional prognostic factors (pathological stage, morphology, histological and nuclear grading, hormone receptors) and newly proposed prognostic factors (ploidy, growth factor receptors, alterations of oncogenes or tumour-suppressor genes, proteases, extracellular-matrix-related antigens).

Non-uniform results, sometimes observed for some proliferation indices (Ki-67, flow-cytometric S-phase cell fraction, PCNA), are probably due to the heterogeneity of the case series under investigation, in terms of stage and treatment, but also to the lack of methodological standardization and quality controls. In addition to the activation of quality control programmes (see Section 3.3), such evidence has led to the establishment of guidelines for the comparison of results and for assessing the prognostic relevance of any biological variable (17). Such prerequisites can be summarized as follows:

- local–regional treatment (to prevent the interference by systemic treatments on natural history)
- case series adequate for size and follow-up
- assessment of interlaboratory consistency of the results
- assessment of the relative prognostic contribution by multivariate analysis including pathological and other biological consolidated prognostic factors

The newly proposed proliferation indices can be validated and considered potentially useful for clinical application only when such criteria are fulfilled.

Results from studies on patients with advanced disease and treated with systemic therapies showed that the proliferative fraction eventually emerged as an indicator of long-term clinical outcome, notwithstanding the impact of therapy on the natural history. This finding, which has been reported for stage II and III breast cancer as well as for non-Hodgkin's lymphomas at all stages, indicates that the original biological aggressiveness (reasonably reflected by cell kinetics) generally prevails over treatment response on long-term clinical outcome, independently of other clinicopathological and biological features.

4.2 Response to treatment

In general, no relationship has been observed between the median proliferative fraction specific for the different histotypes and their clinical sensitivity to cytotoxic agents. Unlike the finding in experimental tumours, this excludes (for human tumours) a direct relation between proliferative rate and response to drugs. Natural drug resistance for some tumour types or inappropriate treatments (in terms of type, schedule and intensity) may be equally responsible for the absence of such a relationship. However, emerging evidence from retrospective studies suggests a relationship between proliferative fraction and treatment efficacy in subsets of patients with potentially sensitive tumours given appropriate treatment (16). In fact, intensive chemotherapy regimens (including S-phase-specific agents given at high dosages) appear to be necessary to counteract the aggressive outcome of rapidly proliferating, advanced head and neck cancers, ovarian and breast cancers, whereas less aggressive treatments seem to be sufficient to control the indolent evolution of slowly proliferating tumours. In particular, hormonal treatment appears highly effective for slowly proliferating breast cancers, but inadequate for rapidly proliferating tumours even in the presence of high oestrogen-receptor contents.

All these data have been confirmed in the adjuvant setting on early disease, have been obtained from retrospective correlative analyses, and need to be verified by prospective studies, which are ongoing mainly for breast cancer.

Acknowledgements

The authors wish to thank B. Johnston and R. Vio for editing the manuscript, and S. Veneroni and E. Benini for their helpful advice in drafting the Protocols. This work was supported in part by the Italian Association for Cancer Research (AIRC).

References

1. Tubiana, M. and Courdi, A. (1989). *Radiother. Oncol.*, **15**, 1.
2. Meyer, J. S. (1991). *J. Clin. Immunoassay*, **14**, 164.
3. Silvestrini, R., Daidone, M. G., and Costa, A. (1989). *Tumori*, **75**, 367.
4. Merkel, D. E. and McGuire, W. L. (1990). *Cancer*, **65**, 1194.
5. Quinn, C. M. and Wright, N. A. (1990). *J. Pathol.*, **160**, 93.
6. Slocum, H. K., Pavelic, Z. P., Rustum, Y. M., Creaven, P. J., Kabakousis, C., Takita, H., and Greco, R. (1981). *Cancer Res.*, **41**, 1428.
7. Teodori, L., Trinca, M. L., Salvati, F., Berettoni, L., Storniello, G., and Göhde, W. (1992). *Int. J. Cancer*, **50**, 845.
8. Silvestrini, R., Sanfilippo, O., and Tedesco, G. (1974). *Cancer*, **34**, 1252.
9. Sasaki, K., Ogino, T., and Takahashi, M. (1986). *Stain Technol.*, **61**, 155.
10. Raza, A., Miller, M. A., Mazewski, C., Sheikh, Y., Lampkin, B., Sawaya, R., Crone, K., Berger, T., Reising, J., Gray, J., Khan, S., and Preisler, H. (1991). *Cell Prolif.*, **24**, 113.
11. Gerdes, J., Li, L., Schlueter, C., Duchrow, M., Wohlenberg, C., Gerlach, C., Stahmer, I., Kloth, S., Brandt, E., and Flad, H. D. (1991). *Am. J. Pathol.*, **138**, 867.
12. Cattoretti, G., Becker, M. H. G., Key, G., Duchrow, M., Schlüter, C., Galle, J., and Gerdes, J. (1992). *J. Pathol.*, **168**, 357.
13. Galand, P. and Degraef, C. (1989). *Cell Tissue Kinet.*, **22**, 383.
14. Silvestrini, R. on behalf of the SICCAB Group for quality control of cell kinetic determination (1991). *Cell Prolif.*, **24**, 437.
15. Silvestrini, R. on behalf of the SICCAB Group for quality control of cell kinetic determination (1994). *Commun. Clin. Cytometry*, **18**, 11.
16. Silvestrini, R. (1994). *Cell Prolif.*, **27**, 579.
17. McGuire, W. L. (1991). *J. Natl. Cancer Inst.*, **83**, 154.

5

Chromosome analysis in cell culture

ANWAR N. MOHAMED and SANDRA R. WOLMAN

1. Introduction

The study of chromosome morphology and number (cytogenetics) describes and defines the cellular vehicles of inheritance. Structural features of the entire set of chromosomes within a cell can only be visualized in the late prophase and metaphase stages of cell division (mitosis); normally the mitotic process occupies a relatively brief (30–40 min) fraction of the entire cell cycle. Thus, only actively replicating tissues are amenable to cytogenetic investigation. The few tissues with sufficient spontaneous cell division to permit direct chromosome preparation, without culture *in vitro*, are bone marrow, trophoblast, testis, and some examples of malignant tumours. In order to obtain metaphase chromosome preparations from other tissues, it is necessary to induce cell division. Therefore, the cultivation of these tissues *in vitro* has become a critical component of cytogenetic investigation. It is essential to remember that this is an indirect approach to the tissues intended for analysis and that selection for growth in culture may result in cells that are not fully representative of the original population. Despite this drawback, almost all the cytogenetic information available on human tumours has been derived from cells grown in culture for various time periods. The methods described herein were developed for culture and cytogenetic analysis of human cells; most are applicable to other mammalian cells from the same tissue sources with minor modifications.

In general, chromosome preparation from cell culture consists of three steps. The first step is arrest of the cell cycle in or before metaphase. Several mitotic inhibitory agents are available, the most commonly used being colchicine, its analogue, colcemid, and the plant alkaloids, vincristine and vinblastine. These agents disrupt the mitotic spindle fibres, freeing the chromosomes from the metaphase plate and allowing them to spread. They also cause chromosomal contraction, facilitating identification and analysis. The second critical step is a hypotonic treatment which results in water intake, swelling of the cells, spreading the chromosomes, and rupturing red cells if present. Potassium chloride (KCl) solution at 0.075 M strength is commonly used and appears least damaging to chromosome morphology and structure. Other suitable hypotonic solutions are 1% sodium citrate or serum diluted to 20–30% in distilled water. Fixation is the third step. The usual fixative for chromosome preparation is a freshly made mixture of three parts absolute methanol to one part glacial acetic acid, a modified Carnoy's fixative, that preserves DNA and enhances spreading and flattening of chromosomes.

Several approaches may be used to validate the results of chromosome studies on cultured cells. Indirect confirmation that they are representative of the original tissue can be obtained from cellular morphology, by formation of specialized structures or growth patterns in culture, by histochemical evidence demonstrating similarities of antigenicity, cell secretion, or other immunocytochemically reactive products. In addition, DNA content or ploidy evaluation can provide important information for comparison of cell populations *in vivo* and *in vitro* (1). More direct evidence can now be obtained using *in situ* hybridization techniques to examine interphase cells from the original sample; if the same profile of chromosomes can be illustrated in both original and cultured cells, then the latter studies are likely to be representative. For example, we used centromere-specific chromosome probes to confirm cytogenetic observations on renal tumours in tissue culture by demonstrating similarity of type and extent of aneuploidy in disaggregated cells from the original tumour masses (2).

2. Blood and bone marrow cultures

2.1 Blood

Blood is one of the easiest and the most accessible tissues for cytogenetic analysis. Clinically, it is used for the determination of constitutional karyotype and for assessment of chromosome damage in individuals exposed to environmental hazards. Blood has several advantages over other tissues. Cell cycles are well characterized and can be synchronized for the preparation of elongate chromosomes for high resolution banding. Large numbers of metaphase spreads can be obtained after only 2–3 days in culture. Normal adult blood contains many cell types although most are non-dividing. Red cells and platelets contain no nuclei, and, among the nucleated white cells (lymphocytes, granulocytes, and monocytes), spontaneous cell divisions are rare. The modification that permits ready access to chromosome preparation is exploitation of mitogens that stimulate lymphocytes (T and/or B) to divide. The most commonly used mitogen is phytohaemagglutinin (PHA), which activates a complex pathway involving T-cell antigen receptors, tyrosine kinases, and lymphokines. The predominant growth factor is interleukin-2 (IL-2) which stimulates proliferation. Other lymphocyte mitogens include pokeweed mitogen (PWM), protein A, and lipopolysaccharides (LPS). These mitogens induce lymphocyte divisions for only a few generations. In addition, Epstein–Barr virus (EBV) can be used to produce long-term cultures with indefinite cell proliferation by immortalizing B-cells to lymphoblastoid cell lines.

2.2 Bone marrow optimal culture conditions

Bone marrow cytogenetic studies are of increasing importance in the diagnosis and risk assessment of patients with leukaemias. Several procedures have improved chromosome preparation of cultured bone marrow samples, although direct preparation remains the method of choice in many cytogenetic laboratories because of speed, lesser expense and, in some cases, better reflection of the true karyotypic picture *in vivo*. Most laboratories use short-term culture (18–72 h) routinely to obtain better quality metaphase cells and higher mitotic indices. Moreover, it appears that karyotypic aberrations in some leukaemias, for example, the t(15;17) in acute promyelocytic leukaemia, may not be observed in direct marrow harvest but is evident after 1–3 days of cultivation (3, 4). Neither approach is optimal for detection of tumour metaphases in all cases and, when feasible, application of more than one method yields the greatest diagnostic information. The choice of procedures should be determined by individual laboratory preference based on experience, time pressures, and type and volume of diagnostic work.

Cytogenetic studies of acute lymphocytic leukaemias (ALL) are technically more difficult than those of myeloid leukaemias. The metaphase chromosomes are often spread poorly and display ill-defined morphology with indistinct bands, making accurate analysis difficult. Recent improvements in cell culturing, harvesting, and staining procedures have minimized these problems and led to better identification of chromosome abnormalities (5). Cultures for ALL are brief (6–24 h) since longer-term cultures (48 h or more) tend to favour the growth of cells with a normal karyotype. Chronic lymphoproliferative disorders (LPD) are characterized by malignant proliferation and accumulation of relatively mature lymphocytes in the bone marrow and/or blood. Approximately 90% of all LPD are of B-cell origin and the remaining 10% are T-cell. Chromosome abnormalities in chronic LPD are not well characterized because of low spontaneous proliferation rate, inconsistent response to mitogens, and competitive proliferation of residual normal T- and B-lymphocytes. The introduction of mitogens that primarily stimulate B-cells has improved our understanding of some of these malignancies. The optimal B-cell mitogens are LPS, EBV, protein A, and tetradecanoylphorbol-13-acetate (TPA). PHA and pokeweed mitogens alone or in combination are alternative mitogens for T-cell LPD. Maximum stimulation is usually seen after 4–5 days in culture.

2.3 Methotrexate synchronization

Methotrexate (MTX) synchronization for blood was first described by Yunis in 1978 and then applied to bone marrow culture by Hagemeijer *et al.* (6). This technique involves increasing the proportion of late prophase and early metaphase cells by synchronizing the cell cycle with MTX block, followed by thymidine release. Elongated chromosomes with 500–2000 bands per haploid set can be obtained (7). These methods have led to detection of small chromosomal deletions that define clinical syndromes, such as Miller–Dieker, Prader–Willi, or Angelman's syndromes and are extremely important in precise definition of specific breakpoints (8). Further modification of the technique for bone marrow by Yunis (9) resulted in high quality and finely banded mitotic spreads. Cell-synchronized cultures are reliable in myeloid leukaemias (10, 11). In contrast, in acute lymphocytic leukaemia (ALL), synchronization may interfere with tumour cell growth resulting in a low mitotic index or overgrowth of normal cells. The technique is labour-intensive and requires precise timing (especially the duration of thymidine incubation). Ethidium bromide is an easier alternative method to obtain elongated chromosomes (*Protocol 4*).

Protocol 1. Culturing of blood and bone marrow

A. *Blood culture initiation*

1. Add 0.8–1 ml of whole blood to 10 ml culture medium in a T-25 flask and culture at 37°C in a humidified chamber 5% CO_2. Incubate for either 48 or 72 h.

Note: The most commonly used media for blood cell culture are RPMI-1640, Eagle's MEM, BME 199 or Ham's F12 medium, usually supplemented with 10–20% fetal bovine serum (FBS), 1% L-glutamine, 1% penicillin, 1% streptomycin and 2% PHA. (Special studies for chromosomal fragile site determination or for synchronized cultures require different media or specific additives.)

B. *Bone marrow culture initiation*

1. Put 1–2 ml of bone marrow aspirate in a sterile heparinized container.
2. Wash cells with Hank's solution and then culture in 10 ml of complete bone marrow culture medium (BMCM), containing 80% RPMI 1640, 20% FBS supplemented with 1% penicillin, 1% streptomycin and 1% L-glutamine.
3. Optimum cell density for a bone marrow culture is 10^6 cells/ml. Cultures containing significantly greater numbers of cells are more likely to fail or to yield poor quality metaphases.
4. Incubate cells in T-25 flasks, with the cap loosened, at 37°C, in a 5% CO_2 humidified chamber. Bone marrow cultures appear to grow better if the surface area is large so that flasks should be horizontal.

Protocol 2. Standard chromosome harvesting of blood or bone marrow cultures

1. Add colcemid to a final concentration of 0.05 μg/ml and reincubate for 45 min at 37 °C.
2. Transfer cell suspension to a 15 ml centrifuge tube and centrifuge at 230 *g* for 10 min.
3. Discard the supernatant.
4. Slowly resuspend the pellet by tapping gently with your finger on the centrifuge tube.
5. Add a prewarmed hypotonic solution of 0.075 M KCl slowly with agitation to a final volume of 7 ml.
6. Incubate for 12 min in a water bath at 37 °C.
7. Centrifuge as in step 2 and discard the supernatant.
8. Resuspend the pellet as in step 4 and slowly add a few drops of freshly made fixative, allowing fixative to run down the sides of the tubes. Make the volume up to 10 ml with fixative. Allow the mixture to stand on ice for 1 h.
9. Change the fixative two or more times by repeating steps 7 and 8.
10. Resuspend the pellet in 15–20 drops of fixative (depending on the amount of sediment). The cell suspension should be turbid but not milky.
11. Mix the suspension with a Pasteur pipette. Put two or three drops of cell suspension on to a wet ethanol-cleaned slide.
12. Let the slide dry spontaneously at room temperature.
13. Check the slide under phase-contrast microscopy to determine metaphase quality.

Notes:
Conditions of temperature and especially of humidity are critical for good spreading of chromosomes.
To increase mitotic activity of leukaemic cultures, we routinely perform short-term culture (48–72 h) in the presence of 10% conditioned medium (CM), in parallel with conventional 24–48 h unsupplemented culture. CM can be derived from many sources; we routinely use a human bladder carcinoma cell line (5637) that generates colony-stimulating factors supporting the growth of myeloid stem cells (12). It has resulted in pronounced enhancement of mitotic activity and chromosome morphology even in samples in which the conventional cultures fail to yield mitoses. Moreover, in some cases, minor chromosomally aberrant subclones which are not found in the conventional cultures, are detected in

Protocol 2. *Continued*

cultures supplemented with CM. On the other hand, in a few cases of myelodysplastic syndromes the CM cultures appear to favour normal cell growth (the percentage of abnormal cells is either decreased or totally absent). Therefore, it is important that samples should always be cultured in parallel without supplement, especially if the diagnosis is not certain.

Protocol 3. Methotrexate synchronization method

1. Initiate short-term cultures from bone marrow or peripheral blood as described in *Protocol 1*.
2. After 3–5 h in culture for bone marrow and 48 or 72 h for blood, add methotrexate to a final concentration of 10^{-7} M and reincubate at 37°C for 17 h.
3. Wash the culture twice in 10 ml of Hank's solution to release the cells from the methotrexate block and re-incubate at 37°C with 10^{-5} M thymidine for 6 h.
4. During the last 10 min of incubation, expose the cells to 0.05 µg/ml of colcemid at 37°C and then harvest according to *Protocol 2*.
5. Fix the cells two or three times in freshly made fixative and store at 4°C overnight. The next day, change the fixative twice more. To overcome cell membrane resistance and to maximize spreading and elongation of the chromosomes, drop the cell suspension from a height of 2–3 ft on to a 70% ethanol-cleaned slide placed at a 30° angle. Two to three drops per slide usually produces excellent spreading.

Protocol 4. Ethidium bromide method

1. Initiate short-term cultures from peripheral blood or bone marrow as described in *Protocol 1*.
2. Add colcemid to a final concentration of 0.05 µg/ml.
3. Add at the same time ethidium bromide to a final concentration of 10 µg/ml.
4. Mix well and re-incubate cultures for 90 min at 37°C.
5. Follow steps 2–13 described in *Protocol 2*.

3. Solid tissue cultures

Solid tissues other than haematopoietic must generally be cultured for longer periods (days to weeks) in order to build up a population of dividing cells. The cells most commonly used for prenatal diagnosis are from amniotic fluid and chorionic villi (CVS). The CVS contain spontaneously dividing cells and a direct harvest may yield metaphases. Alternatively, cultivation is usually the method of choice in most prenatal cytogenetic laboratories. Unlike haematopoietic cells, which grow in suspension, the amniotic fluid cells, CVS, skin cells and other solid tissues, including solid tumours, usually grow as a monolayer attached to the vessel surface. Plastic flasks are suitable for the cultivation of most cell types, but one disadvantage of flask culture is that during harvesting the cells have to be dislodged which leads to loss of some cells. When very small tissue samples are available, it is preferable to culture in smaller vessels. Chamber slide and coverslip cultures have smaller surfaces and yield metaphase cells suitable for harvest in shorter times than if samples are grown out in flasks. Harvests *in situ* are generally faster (within 3–7 days at very low cell density) and have the advantage of minimum cell loss. Moreover, this method is very important in solid tumours for demonstration of cytogenetic heterogeneity and clonality.

3.1 Culture conditions of solid tumours

Much less is known about the cytogenetic abnormalities of solid tumours than haematological malignancies, primarily due to the technical difficulties in processing solid tumours (13). Success in obtaining metaphases is dependent on the tumour type, as well as on technical factors such as specimen transport. Direct analysis of solid tissues is usually limited by low rates of cell turnover. In addition, solid tissues require disaggregation into single cell suspensions or small clusters of cells prior to chromosome preparation. Mechanical dissociation, such as mincing and forcing the tissues through needles or wire mesh usually results in reduction of cell viability. The introduction of enzymatic tissue digestion has been an important advance in solid tissue culture and metaphase analysis. Enzymic digestion reduces the extensive mechanical processing and increases the yield of viable suspended cells. The best and the most satisfactory enzyme for tissue disaggregation is collagenase (14). Trypsin digestion is often associated with cell lysis. Different disaggregation protocols have been described (15). It should be noted, however, that loss of aneuploid populations may result from disaggregation procedures, particularly with enzymatic digestion (16).

Contamination and selective growth of normal cells are other obstacles to successful solid tumour tissue culture. Normal diploid karyotypes are often found in solid tumour cultures, but their significance remains unclear. Although they may represent inflammatory or stromal cells, it is difficult, in many instances, to rule out the possibility that the diploid cells represent a component of tumour cell populations. Selection for growth in culture may be affected by the source of the tumour material, the methods of tissue transport and disaggregation, the method of primary culture and the type of culture medium. These factors may eliminate those cells with abnormal karyotypes, while retaining cells with normal karyotypes. When cells are chromosomally abnormal, they are generally accepted as representing tumour cells. Further, as techniques have improved and periods of growth in culture are decreased, the frequency of abnormal metaphase cells usually increases. The concurrent application of additional parameters, for example, histochemical and molecular markers or DNA flow cytometry, may be helpful in assessing the significance of cytogenetic results. For example, in a study of renal tumours, concurrent flow cytometry of several cases indicated that the fresh tumours had populations that were aneuploid by DNA content which were never identified in culture (1).

The proliferation and growth of most solid tumours *in vitro* depends on adherence of cells to a substrate. Adhesion is dependent on the contact between the cells and attachment factors adsorbed to the growth surface. One of the major attachment factors, fibronectin, is provided by serum in the medium. Additional supplementation of medium is needed for epithelial cells, especially when serum-free or low-serum medium is used to deter fibroblast growth, and a wide variety of growth factors, hormones, vitamins, and detoxification reagents have been tested. Flasks may also be coated with collagen or extracellular matrix to enhance attachment. The use of feeder layers, three-dimensional growth in agar, or growth on floating collagen gels are other possibilities. Some tumour cells can be cultured in simple standard medium, e.g. RPMI 1640, Eagle's MEM, Ham's F12, or McCoy's 5A, whereas others require complex, carefully defined medium with multiple supplements. Generally 15–20% FBS, 1–2% L-glutamine, 1% penicillin and 1% streptomycin are added to the basic medium.

Protocol 5. Culture initiation from solid tissues

1. Examine the tissues to assess viability and note any gross evidence of haemorrhage, infarction, or contamination. Dissect any such areas from the sample before processing. Vital dye exclusion is sometimes a useful measure of cell preservation. Mince the sample finely and/or treat with enzyme so that a suspension of single cells and small clumps is achieved.
2. Plate the cells at a final concentration of $1–2 \times 10^6$ cells/5 ml of medium in T-25 tissue culture flasks or $5–10 \times 10^4$ cells/ml in coverslip culture.
3. Incubate the cultures at 37°C in a humidified 5% CO_2 chamber. Do not

disturb the culture for one to two days to permit attachment. Then exchange the culture medium either partially or completely. If the remaining floating cells are suspected of being viable, transfer them to another flask to examine for growth of cells in suspension.

4. Inspect the cultures daily with an inverted microscope for cellular attachment, flattening, and proliferative activity. Also monitor for fibroblast cell growth and harvest the culture for chromosome analysis before these take over the culture. Because of the cellular heterogeneity and asynchronously dividing population within the tumour, multiple harvests after differing time periods are highly recommended.
5. Colonies of cells with differing morphology are found frequently within the same culture; document this by photography. Chamber slide cultures are advantageous for demonstration of cytogenetic heterogeneity and clonality. Localize and photograph a particular colony and then study the same colony after chromosome preparation, to permit correlation of cytogenetic results with cell and colony morphology.

Protocol 6. Chromosome harvesting of solid tissues

A. *Monolayer flask harvest*

Determining the appropriate harvest time is important. First, harvest should be performed when the culture is subconfluent and cells are actively proliferating. If multiple flasks have been set up, the first harvest should be as soon as possible, and the remainder harvested consecutively over several days.

1. Add colcemid to a final concentration of 0.05 μg/ml for 1–3 h depending on relative mitotic activity of the culture, which can often be estimated by the number of refractile, rounded cells.
2. Remove all medium from the flask and collect it in a 15 ml centrifuge tube.
3. Wash the cells gently with 3 ml PBS solution and transfer to the same tube.
4. Add 0.5–1 ml of prewarmed (37°C) trypsin to the flask.
5. Incubate for 3–5 min. Check frequently in an inverted microscope for cell detachment.
6. As soon as detachment is nearly complete, transfer the cell suspension gently to the same centrifuge tube.
7. Follow steps 2–13 described in *Protocol 2*.

Protocol 6. *Continued*

B. In situ *harvest*

Cultures should be harvested when colonies are still small and have actively dividing cells. Overgrowth of cultures will have an adverse effect on chromosome spreading.

1. Add colcemid to a final concentration of 0.05 μg/ml and incubate for 2 h.
2. Aspirate the medium and add 2 ml of 0.75% Na citrate for 30 min.
3. Slowly and gently add 2 ml of freshly made fixative down the side of the dish.
4. After 2 min, aspirate the hypotonic fixative mixture and replace with 2 ml of fixative. Leave it for 20 min.
5. Change fixative twice interrupted by a 20 min rest period.
6. Aspirate all fixative and let the coverslip dry completely.
7. Mount the coverslip, cell side upward, on to a labelled slide with mounting medium such as pro-texx (Baxter Diagnostic Inc.).
8. Check the slide by phase microscopy for mitotic index and chromosome spreading.

Note: Check humidity in harvest room. It should be between 55 and 57%.

4. Staining and banding methods

Several staining methods allow individual identification of chromosomes by differences in size and internal structure. Originally, stains such as orcein or Giemsa were used; they give a uniform unbanded appearance to the chromosomes (solid stain), but are useful for studies of chromosome breakage and for precise counting of double minutes (small, non-centromeric chromosome fragments). They have been largely replaced by techniques which produce differential staining along the chromosomes (banding). The major banding techniques, G- (trypsin-Giemsa), Q- (quinacrine), or R- (reverse) banding are all suitable for routine use and all display the same underlying unique structural chromosome patterns. Many involve a denaturing step, employing either enzymic digestion, or concentrated salt solution, to partially digest chromosomal proteins prior to staining.

Other techniques are available for special applications and research interests. C-banding, described below, displays heterochromatic regions mainly near the centromeres. G-11 staining, used to identify chromosomes in human/mouse somatic cell hybrids, entails staining with Giemsa at pH 11.0 which differentiates human chromosomes (blue with purplish-red centromeres) from rodent chromosomes (uniformly purple) (17). Silver staining identifies nucleolar organizer regions, located on short arms of the human acrocentric chromosomes (nos. 13, 14, 15, 21, and 22); the patterns are polymorphic (individually different) and heritable (18). DAPI (4′,6-diamidino-2-phenylindole)/Distamycin A staining highlights specific regions on chromosomes 1, 9, 16, the proximal short arm of chromosome 15, and the distal long arm of the Y (19). It is useful in identifying markers, variants and translocations of chromosomes 15 and Y.

4.1 Trypsin–Giemsa staining (G-banding)

This most widely used technique for routine chromosome analysis uses trypsin to denature chromosomal proteins, followed by staining with Giemsa. Each chromosome pair shows a unique pattern of light and dark bands, which allows unequivocal definition of each chromosome and subregions within it.

Protocol 7. G-banding

1. Age the slides for 2 h in a dry oven at 70°C.
2. Prepare solutions for the series of coplin jars shown below.
3. Incubate slides in each coplin jar in the indicated sequence.

Jar	*Contents*	*Time*
1	3.5 ml Difco trypsin 46.5 ml pH 7.0 buffer	20–60 sec
2	50 ml pH 7.0 buffer	15 sec agitation
3	50 ml pH 7.0 buffer	30 sec agitation
4	48.5 ml pH 6.8 buffer, 1.5 ml Giemsa	3–5 min
5	DI water	Rinse well and air dry

Notes:

- Gurr's buffer tablets and Gurr's Giemsa stain are commercially available from Bio/Medical Specialties.
- Optimal time for each batch of trypsin and Giemsa should be determined on a test slide.

4.2 Quinacrine staining (Q-banding)

This method requires examination by fluorescence microscope. The specific pattern of bright Q-bands corresponds almost exactly to dark G-bands. Q-banding is quick, reliable, and does not require aged or pre-treated slides. It is informative with poor preparations as well as with old slides.

Protocol 8. Q-banding

Solution

- Quinacrine stain: Dissolve 30 mg Quinacrine dihydrochloride in 50 ml McIlvaine's buffer pH 5.8. Filter and store the stain at 4°C in the dark by wrapping the jar with aluminium foil. Change the stain every two weeks. Quinacrine dihydrochloride (atebrin) is available from Sigma.
- McIlvaine's buffer (pH 5.8): dilute 480 ml of 0.5 M Na_2HPO_4 and 160 ml of 0.5 M citric acid to 2000 ml with deionized (DI) distilled water. Check pH prior to use. Keep solution in the refrigerator.

1. Prepare solutions for the series of coplin jars shown below.
2. Incubate slides in each coplin jar in the indicated sequence.

Jar	*Contents*	*Time*
1	95% ethanol	5 min
2	Quinacrine stain	20 min
3	DI water	few dips
4	DI water	few dips
5	McIlvaine's buffer	rinse well

3. Mount slides in McIlvaine's buffer with no. 1 coverslip. Blot excess buffer and seal edges with rubber cement. Store stained slides in the dark in refrigerator until ready to analyse.

4.3 Reverse (R-banding)

This stain is useful for analysis of deletions or translocations involving the chromosome ends (telomeres). R-banding is the routine method of cytogenetic analysis in many European laboratories. The stain for R-banding by fluorescence microscopy is Acridine Orange and the alternative method for standard light microscopy uses heat and Giemsa. The resulting chromosome pattern displays the reverse of trypsin–Giemsa (pale G- or Q-bands = darkly stained or brightly fluorescent R-bands).

Protocol 9. R-banding by Acridine Orange

1. Incubate one slide at a time in fresh phosphate buffer (pH 6.5) at 85°C for 15–18 min.
2. Remove slide from the buffer and place in 0.01% phosphate-buffered Acridine Orange for 3–5 min.
3. Rinse the slide for 1.5–3 min in phosphate buffer and mount with the same buffer with no 1 coverslip. Press gently to remove excess.

Phosphate buffer (pH 6.5) 32 ml of 0.07 M disodium hydrogen phosphate dodecahydrate ($Na_2HPO_4.12H_2O$) and 68 ml of 0.07 M potassium dihydrogen phosphate (KH_2PO_4). Adjust by adding 0.07 M $Na_2HPO_4.12H_2O$ to the solution.

Protocol 10. R-banding by Giemsa

1. Incubate slide in EBSS at 85°C for 10–15 min.
2. Rinse in cool tap water.
3. Stain with 2% Giemsa for 10–20 min.
4. Rinse in tap water and air dry.

EBSS buffer (pH 6.8) = 10 ml Earle's balanced salt solution (EBSS) (10×) + 0.1 ml 7.5% Na bicarbonate + 89.9 ml distilled water.

Note: Slides can be stained sequentially for both Q- and R-banding with or without removing quinacrine. Usually better R-banding is obtained after Q-banding than with R-banding alone.

4.4 Constitutive heterochromatin banding (C-banding)

C-banding stains the centromeric region of each chromosome and other regions containing heterochromatin, e.g., 1qh, 9qh, 16qh and Yqh. It is used primarily to identify polymorphic variants and the common inversion rearrangement of chromosome 9. It may be helpful for identification of supernumerary or marker chromosomes.

Protocol 11. C-banding

1. Prepare solutions for the series of coplin jars shown below.
2. Incubate slides in each coplin jar in the indicated sequence.

Jar	*Contents*	*Time*
1	0.2 M HCl	30 min
2	DI water	rinse well
3	0.07 N Ba $(OH)_2$	7 min at 37°C
4	DI water	rinse well
5	2 × SSC (saline–sodium citrate)	1 hour at 65°C
6	DI water	rinse well
7	4% Giemsa stain	20 min
8	DI water	rinse well and air dry

5. Nomenclature and interpretation

A metaphase chromosome consists of two chromatids held together at the centromere, which divides the chromosomes into two arms termed the short (p) arm and the long (q) arm. Chromosomes are described on the basis of centromere location as follows: if the centromere is mid-length, the chromosome is metacentric; if the centromere is near but not at the mid-point it is submetacentric; if the centromere is near the end it is called acrocentric, and a centromere at the end is characteristic of a telomeric chromosome (common in rodents, but not normally found in the human karyotype). Before the advent of banding techniques, chromosome size and centromere position were the only distinguishing features for karyotypic analysis. Normal human somatic tissues contain the diploid number of 46 chromosomes, which are separable into 23 pairs, 22 pairs of autosomes and one pair of sex chromosomes. The homologous pairs (all autosomes and the X chromosomes) are matched with respect to banding pattern and order of genetic loci. The sex chromosomes of the male, an unmatched X and Y, share homologous sequences in small regions of their short arms.

5.1 Numerical and structural aberrations

An orderly arrangement of chromosomes from a somatic cell, arranged according to size, centromere position, and banding pattern is identified as a karyotype. The karyotype is described according to the International System for Human Cytogenetic Nomenclature (ISCN); the chromosome number is followed by the sex chromosome complement, e.g., 46,XX or 46,XY for normal female and normal male, respectively. Abnormalities may be either structural or numerical and may involve autosomes, sex chromosomes or both. Numerical anomalies may involve loss or duplication of the entire set (haploidy, tetraploidy), but when an abnormal number results from loss or gain of individual chromosomes it is referred to as aneuploidy. Aneuploidy usually results from non-disjunction (failure of chromosome segregation during cell division) and can occur during meiosis or mitosis. Such errors can produce daughter cells with one copy (monosomy) or three copies (trisomy) of a specific chromosome, instead of the normal two copies (disomy). When a cell has the normal chromosome number, but abnormal numbers (not paired) of individual chromosomes and/or structural aberrations, the cell is called pseudodiploid.

Structural rearrangements result from chromosome breakage, followed by abnormal reunion. The common aberrations are: (i) translocation (balanced and unbalanced, reciprocal and Robertsonian); (ii) deletion (terminal and interstitial); (iii) inversion (paracentric and pericentric); (iv) duplication; and (v) insertion. Less common structural aberrations include isochromosomes, dicentrics, rings, markers, double minutes, homogeneously staining regions and fragile sites. Definitions and detailed descriptions are found in the ISCN (1985) and one should not attempt to analyse a karyotype without this reference (20). These cytogenetic alterations can result in cells with balanced or unbalanced genomes. Trisomies, monosomies, deletions and duplications are usually associated with chromosomal and gene imbalance. Translocations and inversions do not necessarily produce imbalance, but may modify the expression of rearranged genes.

5.2 Tumour cell populations

Tumour cells often display acquired chromosomal aberrations that are not present in normal somatic cells and that serve as markers to identify tumour cells. To the extent that they correlate with tumour morphology or subclassification, they contribute diagnostic information (12). Some tumour cell populations show mixtures of normal and aberrant metaphase cells. The aim of cytogenetic analysis is detection of clonal aberrations. Because each tumour is thought to arise from a single progenitor cell, the existence of several or many cells with the same aberration denotes expansion from that progenitor (clonal proliferation). For this reason, prolonged growth in culture prior to chromosome analysis can result in misleading information and interpretations. Cytogenetically, an abnormal clone is defined when at least two or more cells show the same structural aberration or the same additional chromosome (usually based on examination of 20–30 cells). If cells are hypodiploid, three or more cells must show loss of the same chromosome in order to define a clone, because random, artefactual loss is common.

Tumours are also characterized by heterogeneity of chromosome aberrations from cell to cell. The main line is the largest or dominant clone. This is usually the most frequent chromosome pattern in a tumour cell population but does not necessarily indicate the basic population in terms of tumour evolution. The stemline is the basic abnormal clone of a tumour cell population, assuming that cells evolve from a normal diploid pattern to one of increasing deviation from the norm. Sidelines share the basic chromosomal aberrations seen in the stemline, but they acquire additional changes unique to the individual sideline. Other more complex definitions and nomenclature relating to tumour cell populations are described in Guidelines for Cancer Cytogenetics, a supplement to ISCN, edited by Mitelman (1991) (20).

6. *In situ* hybridization and interphase cytogenetics

The recent development of tools for non-isotopic *in situ* hybridization of chromosomes now permits study of previously inaccessible populations of interphase cells as well as better definition of chromosome rearrangements. Probes are constructed from modified bases introduced into DNA that react with fluorescent or histochemical labels. Because the satellite, highly repetitive DNAs of centromeric regions are unique to individual chromosomes, human centromeres are excellent targets for chromosome enumeration. The number of target sites per nucleus corresponds to the copy number of the particular chromosome. Smaller, unique DNA sequences are recognized either by tagging of several geographically clustered sequences or by inclusion of the segment of interest in a larger vector (cosmid) to increase the target size. Whole chromosome 'painting' is achieved by labelling of hundreds of unique-sequence probes from the entire length of a single chromosome. When used together with an appropriate linked fluorescent label, then the entire chromosome appears painted with the label. These probes are valuable tools for definition of the components of unknown, marker chromosomes.

Fluorescent labels for *in situ* hybridization are preferred for several reasons; reading and interpretation are quite consistent, and the probes are amenable to double-labelling and signal amplification; however, because of fading, the use of fluorescence *ir situ* hybridization (FISH) requires photographic records. Fluorescence microscopy of high quality and special filters are necessary for analysis. Directly labelled probes are available but usually result in considerably weaker signals than the indirect biotinylated or digoxigenin-linked probes which provide options for amplification of the fluorescence signal. Because of fading, it is important to read slides soon after preparation. However, fixed cell pellets and other samples may be held for months, and we have had excellent results on paraffin sections from 20-year-old blocks.

FISH probes are used on metaphase preparations to resolve components of rearranged chromosomes and to detect micro-deletions. FISH is also useful for gene localization, since signals are often present on two, three or even four chromatids of the relevant chromosome, and this striking degree of labelling is not observed with other methods. In interphase cells, FISH-based probes can be used to detect viral sequences, to identify constitutional chromosome anomalies in prenatal specimens, to demonstrate tumour-specific chromosome translocations (using flanking markers tagged with different fluorescent labels), and to evaluate gene amplification. The identification of increases of chromosome number with several centromeric probes simultaneously also provides a rough measure of increased ploidy. This approach has the greatest impact on analysis of interphase nuclei because of overall sample representation and absence of selection for the dividing component of cell populations.

The point of reference for attributing significance to the cytogenetic observations in cultured cells must be their similarities to those of the original tissue source. One can now return to the original specimen, either as a reserved aliquot of the plated disaggregated cells, in sections from the original tissue blocks (21, 22), or as disaggregated cells retrieved from paraffin sections (2). Comparisons with metaphase analysis with respect to gains or losses of whole chromosomes or of specific loci should reveal the extent to which the cultured cells are reflective of the original tissue and the extent to which selection and evolution in culture have modified the results. Examination of large numbers of interphase cells is relatively simple. This type of study is of particular value for epithelial tumours, such as breast or prostate, that have not been informative by conventional cytogenetic assessment because direct harvests have had poor success rates and cultures have often yielded diploid cells.

Protocol 12. Basic steps in FISH

Detailed instructions from the probe manufacturers should be followed for the individual probes.

1. Basic hybridization entails a preliminary exposure to RNase followed by washing in SSC and ethanol dehydration.
2. Denature the prewarmed slides briefly in 70% formamide and 2× SSC with heat.
3. Streak the heat-denatured probe mix on to the slide and seal; incubate slides for several hours.
4. After hybridization, soak slides and expose again to a formamide/SSC mixture.
5. Accomplish specific fluorescent labelling by preblocking, usually with a buffered solution of non-fat dry milk, followed by incubation with the fluorescent reagent, washing, re-blocking with buffered serum, and finally, staining with a fluorescent counterstain for viewing. The entire procedure can be accomplished in 4–24 h.

FISH probes have certain inherent problems. The number of repeats in pericentromeric regions varies with the individual chromosome; if low, the signal will be correspondingly weaker. Some probes cross-react with other centromeric regions or with minor hybridization sites, increasing the apparent number of signals. Loss of target DNA, poor penetration of probe, incomplete or inefficient hybridization, or signal overlap reduce the apparent number of signals. Examination of large numbers of cells, careful determination of suitable internal and external controls, and relatively sophisticated statistical analyses are sometimes necessary to achieve meaningful results.

Central issues in tumour biology amenable to FISH analysis include appreciation of genetic heterogeneity, characterization of premalignant lesions, and better understanding of relationships between primary tumours and their metastases. As more of the human genome is defined, the tools for FISH will expand and the precision of many diagnoses will be refined. The appropriate development and application of these probes will be greatly influenced by results of cytogenetic analysis on cultured cells. These probes when applied simultaneously with markers for cell differentiation, cell turnover, or even specific markers of malignancy, will in turn provide a far greater understanding of the identity, state of differentiation and relatedness of cells in culture to their tissues of origin.

References

1. Wolman, S. R., Camuto, P. M., Schinella, R., and Golimbu, M. (1988). *Cancer Res.*, **48**, 2890.
2. Wolman, S. R., Waldman, F. M., and Balazs, M. (1993). *Genes, Chromosomes Cancer,* **6**, 17.
3. Dewald, G. W., Broderick, D. J., Tom, W. W., Hagstrom, J. E., and Pierre, R. V. (1985). *Cancer Genet. Cytogenet.*, **18**, 1.
4. Knuutila, S., Vuopio, P., Elonen, E., Siimes, M., Kovanen, R., Borgstrom, G. H., and de la Chapelle, A. (1981). *Blood,* **58**, 369.
5. Williams, D. L., Harris, A., Williams, K. J., Brosius, M. I., and Lemonds, W. (1984). *Cancer Genet. Cytogenet.*, **13**, 239.
6. Hagemeijer, A., Smit, E. M. E., and Bootsma, D. (1979). *Cytogenet. Cell Genet.*, **23**, 208.
7. Yunis, J. (1981). *Hum. Genet.*, **56**, 293.
8. Ledbetter, D., Riccardi, V., Airhart, S., Strobel, R., Keenan, B., and Grawford, J. (1981). *N. Engl. J. Med.*, **304**, 325.
9. Yunis, J. (1981). *Hum. Pathol.*, **12**, 540.
10. Yunis, J., Bloomfield, C. D., and Ensrud, K. (1981). *N. Engl. J. Med.*, **305**, 135.
11. Mohamed, A., Clarkson, B., and Chaganti, R. S. K. (1986). *Cancer Genet. Cytogenet.*, **20**, 209.
12. Wolman, S. and Mohamed, A. (1991). *Biochemical and Medical Aspects of Selected Cancers*, Vol. 1, pp. 393–426. Academic Press, London.
13. Sandberg, A. A., Turc-Carel, C., and Gemmill, R. M. (1988). *Cancer Res.*, **48**, 1049.
14. Limon, J., Dal Cin, P., and Sandberg, A. A. (1986). *Cancer Genet. Cytogenet.*, **23**, 305.
15. Trent, J., Crickard, K., Gibas, Z., Goodacre, A., Pathak, S., Sandberg, A. A., Thompson, F., Whang-Peng, J., and Wolman, S. (1986). *Cancer Genet. Cytogenet.*, **19**, 57.
16. Costa, A., Silvestrini, R., and Del Bino, G. (1987). *Cell Tissue Kinet.*, **20**, 171.
17. Bobrow, M. and Cross, J. (1974). *Nature*, **251**, 77.
18. Howell, W. M. and Black, D. A. (1980). *Experientia*, **36**, 1014.
19. Schweizer, D., Ambros, P., and Andrle, M. (1978). *Exp. Cell Res.*, **111**, 327.
20. ISCN (1985). International System for Human Cytogenetic Nomenclature, March of Dimes Birth Defects Foundation and Cytogenetics and Cell Genetics, S. Karger, Basel, Switzerland. (Guidelines for Cancer Cytogenetics (ed. F. Mitelman, Supplement 1991.)
21. Macoska, J., Micale, M., Sakr, W., Benson, P., and Wolman, S. (1993). *Genes, Chromosomes Cancer*, **6**, 17.
22. Micale, M., Sanford, J., Powell, I., Sakr, W., and Wolman, S. (1993). *Cancer Genet. Cytogenet.*, **69**, 7.

6

Assessment of DNA damage in mammalian cells by DNA filter elution methodology

RICHARD BERTRAND and YVES POMMIER

1. Introduction

A variety of DNA lesions can be measured in mammalian cells by the DNA filter elution methodology. All DNA elution methods were developed in Dr Kurt W. Kohn's laboratory (Laboratory of Molecular Pharmacology, NCI, NIH, Bethesda, MD). DNA elution is commonly used to study the effects and mechanisms of action of chemotherapeutic drugs and carcinogens (1–3). The basic DNA elution filter methods were originally designed to assay DNA damage in intact cells or tissues from living animals (1). More recently, the DNA elution filter assays were adapted to study drug mechanisms in isolated nuclei (4–8) and in a reconstituted cell-free system (9–12). Altogether, the various elution methods are currently applied to study the DNA effects of a variety of anticancer agents (topoisomerase inhibitors, alkylating and cross-linking drugs) in cells in culture and to analyse the DNA fragmentation associated with programmed cell death (apoptosis) (9–12).

This chapter will review briefly the types of DNA damage that can be assayed by DNA filter elution and the fundamental concepts underlying the DNA filter elution methodology. The experimental procedures of the various DNA filter elution assays are given in detail. The theoretical aspects of the DNA filter elution methodology are outside the scope of this chapter and have been extensively reviewed recently (1).

2. DNA lesions measured by filter elution assays

The more common types of DNA lesions that can be evaluated by DNA filter elution include DNA single-strand breaks (SSB) (1, 2, 13) as well as double-strand breaks (DSB) (1, 2). SSB and DSB are either protein-associated (PASB) or protein-free ('frank breaks,' FB), (1, 2, 13, 14), DNA–protein crosslinks (DPC) (1, 2, 14), interstrand DNA crosslinks (ISC) (1, 2, 15),

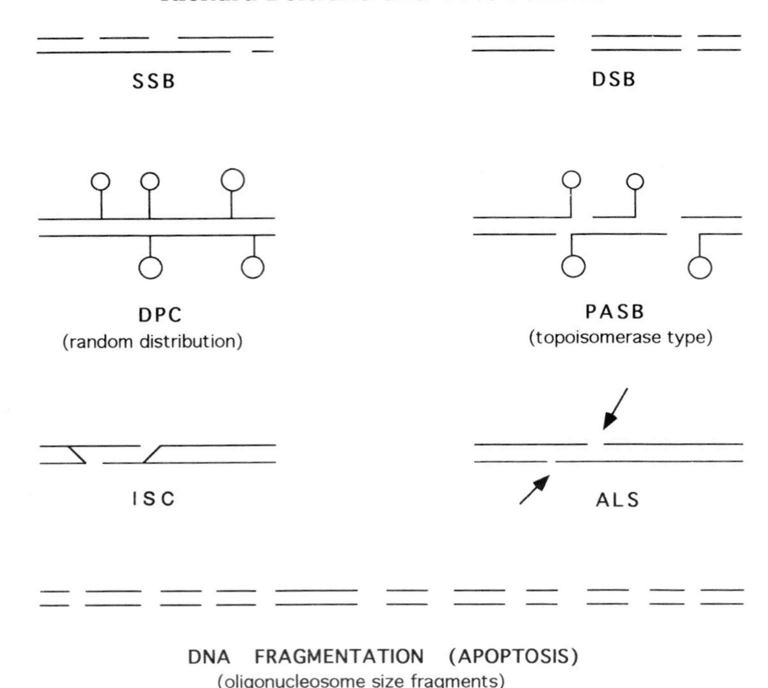

Figure 1. Schematic representation of the various types of DNA lesions that can be detected by DNA filter elution. SSB, single-strand breaks; DSB, double-strand breaks; DPC, DNA–protein crosslinks; PASB, protein-associated strand breaks; ISC, interstrand DNA crosslinks; ALS, alkali-labile sites; apoptosis-associated DNA fragmentation.

alkali-labile sites (ALS) (1, 2, 16) and more recently apoptosis-associated DNA fragmentation (9–12) (*Figure 1*). *Table 1* shows a comparison of the main DNA filter elution methods with typical examples of inducing agents.

3. Basic principles of the DNA filter elution methodology

The principles of the DNA filter elution methods have been reviewed recently (1). The technique uses membrane filters that do not adsorb DNA. Cells are deposited on to a membrane filter and lysed with detergent. The elution solution is then pumped slowly through the filter. The long DNA strands are mechanically retained on the filter whereas short DNA strands elute from the filter as a result of DNA breaks. The procedure avoids extensive sample handling and thus eliminates mechanical damage to DNA (1, 2).

Different filters, lysis, and eluting solutions are used in order to maximize

Table 1. Overview of the DNA filter elution assays

Elution assay	Filter	Lysis solution	Eluting solution	Example
SSB	PC	SDS-ProK	pH 12.1 + SDS	Topo I-II
DSB	PC	SDS-ProK	pH 9.6 + SDS	Topo II
PASB/FB	PVC	LS-10	pH 12.1	Topo I-II/X-rays
DPC	PVC	LS-10	pH 12.1	CDDP
ISC	PC	SDS-ProK	pH 12.1 + SDS	CDDP, HN_2
ALS	PC	SDS-ProK	pH > 12.1 + SDS	BrdUrd, MN
DNA-F	PVC	LS-10	None	Apoptosis

ALS, alkali-labile sites; BrdUrd, bromodeoxyuridine; CDDP, cis chlorodiaminoplatinum DNA-F, DNA fragmentation; DSB, double-strand breaks; DPC, DNA–protein crosslinks; FB, frank breaks; ISC, interstrand DNA crosslinks; HN_2, nitrogen mustard; LS-10, 0.2% sarkosyl, 0.04 M Na_2EDTA, 2 M NaCl (pH 10.0); MN, methylnitrosourea; PASB, protein-associated strand breaks; PC, polycarbonate membrane filter; PVC, polyvinyl (or polyvinyl/acrylic copolymer) membrane filter; SDS-ProK, 2% SDS, 0.025 Na_2EDTA, 0.5 mg/ml proteinase K (pH 10.0); SSB, single-strand breaks; Topo I-II, topoisomerase I and II inhibitors.

or minimize protein adsorption to the filters. This allows the differentiation of protein-associated and protein-free DNA strand breaks (9, 13, 17).

The elution rate can be adjusted to optimize the quantification of the DNA lesions. The elution rate is usually 0.03–0.04 ml/min with fractions collected at 3 h intervals for 15 h for SSB, DSB, PASB, FB, DPC, ISC, and ALS (see *Tables 1* and *2*) (2, 7). Faster elution rates (0.12–0.16 ml/min with 5 or 10 min fractions collected over 30 or 60 min) can be used for SSB above 1000 rad-equivalents (13) and DSB above 10000 rad-equivalents (18).

The pH of the elution solution is above pH 12.0 to assay for DNA single-strand breaks (DNA denaturing or 'alkaline' elution), or below pH 10.0 to assay DNA double-strand breaks (non-DNA denaturing or 'neutral' elution). Alkali-labile sites are detected by their conversion into single-strand breaks using an eluting solution at higher pH such as 12.6–12.8 (19, 20).

DNA–protein crosslinks and interstrand crosslinks are measured as a reduction of the elution rate of DNA single-strand breaks that are generally produced by ionizing radiation. Thus, these DNA filter elution assays include X- or gamma-irradiation of drug-treated cells to introduce an appropriate frequency of single-strand breaks that are needed to detect and measure the relative DNA–protein crosslink and interstrand crosslink frequencies (1, 2).

The quantitative determination of a given DNA lesion is based on calibration and normalization with irradiated cells to produce a random distribution of DNA strand breaks (see Section 6). Internal standard cells are used to define a corrected elution time scale to normalize the elution curves of different samples processed during the same experiment. Calibration standard elution curves are used to convert the elution rate constant of the experimental cells into equivalent frequencies of DNA strand breaks induced by

X- or gamma-ray dose alone. By using this type of calibration standard as reference, DNA lesion frequencies are expressed in rad-equivalents (1, 2). In the case of SSB, one rad-equivalent corresponds to approximately 1 SSB per 10^9 nucleotides. One rad-equivalent DSB corresponds to approximately 0.05 DSB per 10^9 nucleotides (7, 21).

4. Overview of the DNA filter elution procedures

This section describes the basic equipment and methods needed to perform the various DNA filter elution assays. Specific conditions of the various DNA filter elution assays will be described in Section 5. Computations are explained in Section 6.

4.1 Equipment

The basic equipment is as described in Kohn *et al.* (2).

Filtration funnels consist of 25-mm polyethylene filter holder (Swinnex, Millipore Corp.) connected and cemented with epoxy to a 50-ml polyethylene syringe (Swinnex funnel). The syringe orifice should be enlarged with a 0.125 in (3.2 mm) diameter drill to facilitate filling the upper section of the filter without trapping air. The bottom of the filter holder is connected to a 15 gauge stainless-steel needle inserted through a rubber stopper to fit a filtration funnel holder (*Figure 2*) (2).

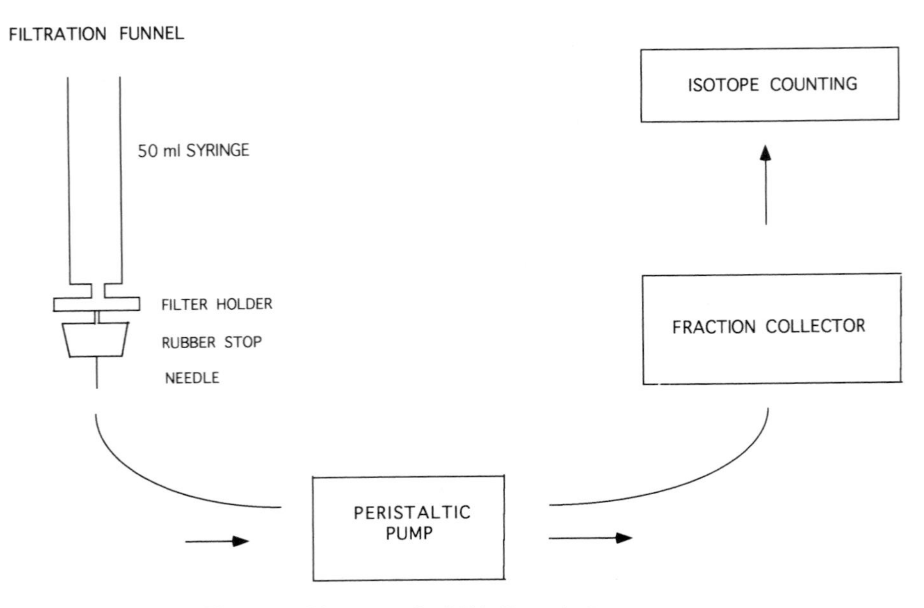

Figure 2. Diagram of a DNA filter elution set up.

The outflow from the elution funnels must be pumped at slow and constant speed without pulsation. We use a Gilson Mini-pulse II 8 channel pump equipped with a slow-speed gearbox (Gilson Instrument Co.). The tubing used for the Gilson peristaltic pumps are Silicon pump tubing (size: I.D. 0.035) from Elkay Products Inc. The filtration funnels are connected to a fraction collector via 0.86 mm (i.d.) polyethylene tubing. Fractions are collected directly into scintillation bottles at intervals of 5 or 10 min (0.12–0.16 ml/min; *short elution* methods) (18, 22) or 180 min (0.03–0.04 ml/min; *long elution* methods) (1) (*Table 2*).

Access to an X-ray or gamma-ray irradiator may pose a problem to some laboratories. Although irradiation at 0°C is the most convenient way to introduce random DNA strand breaks, an alternative method has been described using hydrogen peroxide (23). Similarly, DNA radiolabelling is the most sensitive method for DNA filter elution assays although elutions have been done using fluorometric DNA assays (24).

4.2 DNA labelling and preparation of experimental cell cultures

The DNA of exponentially growing cells is labelled with [^{14}C]thymidine (0.01–0.02 μCi/ml) for approximately one doubling-time and then post-incubated for at least 4 h in isotope-free medium to chase radioactivity into high-molecular-weight DNA (2).

Control and drug-treated monolayer cells are detached by gentle scraping with a rubber 'policeman' and dispersed by repeated pipetting in their medium. Suspension cells are dispersed by repeated pipetting in their medium. Typically 0.5–1.0 $\times$ 10^6 cells (corresponding to at least 5000 d.p.m.) are diluted in 10 ml of ice-cold Hank's balanced salt solution (containing the same drug concentration in the case of cold-reversible lesions, such as topoisomerase I-cleavable complexes induced by camptothecin) (25).

4.3 Calibration and internal standard cells

Calibration standard cells are irradiated on ice with 3 Gy for SSB long methods and 10–20 Gy for SSB short methods. Higher irradiation doses are necessary for DSB calibrators because ionizing radiation produces markedly less DSB than true SSB (approximately 20-fold (21)). For the DSB long method the X-ray doses range from 30 to 100 Gy; for the DSB short method, the X-ray doses range from 100 to 300 Gy. The calibration standard elution curves are used to convert the elution rate constants of experimental cells into rad-equivalents (or Gy-equivalents; 1 Gy = 100 rads (for calculations see Section 6)).

The precision of elution assays can be greatly increased by adding internal standard cells to define a corrected time scale and to normalize the experi-

Table 2. Schematic flowchart of the different filter elution assays

Elution method		SSB long method	SSB short method	DSB long method	DSB short method	ISC	PASB FB	DPC
Cell preparation	Experimental cells	none	none	none	none	3 Gy	none	30 Gy
	Calibrator cells	3 Gy	20 Gy	30–100 Gy	100–300 Gy	3 Gy	3 Gy	30 Gy
	Internal Std cells	3 Gy	20 Gy	30–100 Gy	100–300 Gy	3 Gy	3 Gy	30 Gy
Type of filter		PC	PC	PC	PC	PC	PVC	PVC
Cell deposition on to filter								
Cell lysis before elution	Lysis solution	5 ml SDS	5 ml SDS/ProK	none	none	5 ml SDS/ProK	5 ml LS-10	5 ml LS-10
	Wash solution	none	5 ml EDTA			none	10 ml EDTA	10 ml EDTA
Connect elution tubing (Pump off)								
Elution solution	Loading with micropipette	2 ml SDS/ProK 1 ml Pr4/SDS, pH 12.1	3 ml Pr4/SDS, pH12.1	2 ml SDS/ProK 1 ml Pr4/SDS, pH 9.6	2 ml SDS/ProK 1 ml Pr4/SDS, pH 9.6	2 ml SDS/ProK 1 ml Pr4/SDS, pH 12.1	3 ml Pr4,[a] pH 12.1	3 ml Pr4, pH 12.1
	Funnel loading	40 ml Pr4/SDS, pH 12.1	10 ml Pr4/SDS, pH 12.1	40 ml Pr4/SDS, pH 9.6	10 ml Pr4/SDS, pH 9.6	40 ml Pr4/SDS, pH 12.1	40 ml Pr4, pH 12.1	40 ml Pr4, pH 12.1
Pump speed		0.035 ml/min	0.14 ml/min	0.035 ml/min	0.14 ml/min	0.035 ml/min	0.035 ml/min	0.035 ml/min
Fractions	Time interval	3 h	5 min	3 h	10 min	3 h	3 h	3 h
	Number	5 fractions	6 fractions	5 fractions	6 fractions	5 fractions	5 fractions	5 fractions

This table is organized chronologically with respect to a typical alkaline elution. It describes from top to bottom (see left column) the preparation of the cells, the type of filter used for each elution, the cell deposition step on to the filter, the cell lysis, followed by the elution tubing connection, the addition of lysis or elution solution, the pump speed for each type of elution, as well as the number of fractions and time interval between fractions.
Solutions and types of filter are described in Section 5.1.
[a] Pr4: tetrapropyl ammonium hydroxide elution solution without SDS.

mental elution curves (1, 2). The use of internal standard cells limits the variability between different lines (funnels) in a given experiment.

Internal standard cells are generally labelled with [^{3}H]thymidine (0.02–0.2 μCi/ml) for 16 h and chased for at least 4 h in isotope-free medium.

The [^{3}H]thymidine labelled cells are irradiated on ice with 3 Gy for SSB long method and ISC assays, with 10 or 20 Gy for SSB short method, and with 30 Gy for DPC assays (see *Table 2*).

An aliquot of these irradiated internal standard cells (at least 10 000 d.p.m.) is then mixed with experimental cells labelled with [^{14}C]thymidine before the elution filter assay. The total number of cells per filter should not exceed 10^6. The elution curves of experimental cells will be plotted versus the elutions of internal standard cells (for calculations see Section 6).

Internal standard cells are usually not used in the case of DSB assays where the total number of cells per filter is limited. Also, in the case of the topoisomerase I inhibitors, camptothecin, it is difficult to use internal standard cells because camptothecin induces topoisomerase I-cleavable complexes (i.e. SSB and DPC) at ice temperature in the internal standard cells (25).

4.4 Cell lysis, DNA elution procedures, and treatment of filters

Experimental cells (control, calibration standard and treated-samples to be analysed) that have been mixed with the internal standard cells (total number of cells $<10^6$) are kept on ice until their loading into the elution funnel. It is recommended that the elution is primed by pipetting 2–3 ml of ice-cold balanced salt solution (such as Hank's solution or PBS) directly into the hole of the Swinnex filter holder using a micropipette. Care should be taken not to puncture the filter. The cell suspension is then loaded into the funnel and the solution is allowed to flow by gravity.

The lysis solution is added to the funnel immediately after the cells are deposited and the last drop of the cell suspension has dripped through. Delay in adding the lysis solution may minimize DNA lesions that may repair while the cells warm up to room temperature.

After the filtration funnel has been connected to the peristaltic pump, about 3 ml of the appropriate elution solution (see Section 5 and *Table 2*) is added slowly through the filter holder hole with an automatic pipette bearing a narrow tip. The rest of the elution solution is then poured into the funnel reservoir (see *Table 2*). The peristaltic pump is turned on and fractions collected (see *Table 2*).

When all fractions have been collected, the elution solution remaining in the funnel reservoir is discarded. The pump is turned on to maximum speed to empty the lines into a new scintillation bottle. This fraction corresponds to the 'line' (*Table 3*).

The filter is then removed and placed in a scintillation bottle and 0.4 ml of

1 M HCl is added. The vial is sealed and heated for 1 h at 65°C to depurinate the DNA. After removing the vials from the oven, 2.5 ml of 0.4 M NaOH is added and the vials are allowed to stand for 45–60 min at room temperature to release the DNA from the filter into solution and allow maximum counting efficiency. Then 2 ml of H_2O is added to the 'Filter' vial (*Table 3*).

During the filter treatment, the Swinnex is reassembled and 5 ml 0.4 M NaOH is pumped through the tubing to collect the remaining radioactive material. This fraction is collected as the 'Wash' (*Table 3*).

At the end of the elution, an additional 10 ml of H_2O is pumped through the funnel to remove the NaOH from the elution set up. Otherwise, the lines tend to accumulate crystals. As a result, elution rates would become variable and the elution tubing blocked.

All fractions contained in scintillation bottles are collected. In the case of the short methods 4 ml of water is added to each elution fraction to keep the ratio water:liquid scintillation cocktail around 0.5 for optimum counting efficiency. Then 10 ml of scintillation cocktail is added [Aquassure^R (New England Nuclear) to which 0.7% glacial acetic acid has been added to prevent chemiluminescence in the alkaline solutions]. Radioactivity is then measured in each fraction by double-isotope counting (^3H for internal standard cells and ^{14}C for experimental cells) by liquid scintillation spectrometry.

Data are computed as described in Section 6.

5. Specific conditions for each of the DNA filter elution assays

This section deals with the specific conditions for the various DNA filter elution assays (*Tables 1* and *2*). The general procedures have been discussed in Section 4. For elution curve analyses and DNA lesion frequency calculations, the reader should refer to Section 6.

5.1 Filters and solutions

5.1.1 Filters

For SSB, DSB, ISC and ALS, non-protein adsorbent filters are used. Polycarbonate membrane filters 2 μm pore size, 25 mm diameter can be obtained from Nucleopore Corporation or Poretics Corporation.

For DPC, protein-concealment of DNA breaks, and DNA fragmentation assays, protein-adsorbent membrane filters 25 mm diameter can be obtained from:

- Gelman Sciences Inc.: PVC (polyvinyl chloride)/acrylic copolymers, Metricel^R, 0.8 μm pore size
- Poretics Corporation: PVC, 2 μm pore size

5.1.2 Solutions

(i) EDTA washing solution pH 10.0.

(a) Prepare Na_2EDTA (0.1 M) from 37.2 g of Na_2EDTA (ethylenediaminetetraacetic acid disodium salt: dihydrate) and 8 g of NaOH made up to 1 litre with H_2O. Adjust pH to 10.0 with NaOH (5 M).

(b) For 500 ml of 0.02 M EDTA mix 100 ml of Na_2EDTA (0.1 M) and 400 ml of H_2O and adjust the pH to 10.0.

(ii) LS-10 (lysis solution pH 10.0): add 116.8 g NaCl, 400 ml 0.1 M Na_2EDTA, 6.7 ml of 30% Sarkosyl solution and distilled water to a final volume of 1 litre. Adjust to pH 10.0 with NaOH/HCl.

(iii) Nucleus buffer: add 0.136 g KH_2PO_4, 1.016 g $MgCl_2$, 8.76 g NaCl, 0.380 g EGTA and distilled water to a final volume of 1 litre. Adjust the pH to 6.4 with NaOH/HCl. Add DTT (dithiothreitol) to 0.1 mM final concentration immediately before use.

(iv) Nucleus buffer + Triton X100: add 1 ml of Triton X100 (30% stock solution) to 100 ml of nucleus buffer.

(v) Tetrapropyl elution solution without SDS pH 12.1: add 5.8 g ethylenediaminetetra-acetic acid (free acid) (H_4EDTA), 50 g tetrapropyl ammonium hydroxide (Pr_4 NaOH-40%) (RSA Co.) and distilled water to a final volume of 1 litre. Adjust pH to 12.1 using the Tetrapropyl buffering solutions (RSA Co.).

(vi) Tetrapropyl buffering solutions:

(a) Buffering solution pH > 13: add 2.9 g H_4EDTA, 100 g Pr_4NaOH-40% and distilled water to a final volume of 500 ml.

(b) Buffering solution pH ~ 6.6: add 2.9 g H_4EDTA, 12 g Pr_4NaOH-40%, and distilled water to a final volume of 500 ml.

(vii) Tetrapropyl elution solution with SDS pH 12.1: add 5.8 g H_4EDTA, 50 g Pr_4NaOH-40%, 1 g SDS and distilled water to a final volume of 1 litre. Adjust pH to 12.1 using the Tetrapropyl buffering solutions.

(viii) Tetrapropyl elution solution with SDS pH 9.6: add 5.8 H_4EDTA, 32.5 g Pr_4NaOH-40%, 1 g SDS and distilled water to a final volume of 1 litre. Adjust pH to 9.6 using the Tetrapropyl buffering solutions.

(ix) SDS lysis solution: add 7.51 g glycine, 20 g SDS, 9.30 g Na_2EDTA and distilled water to a final volume of 1 litre. Adjust pH to 10.0 with NaOH/HCl.

(x) SDS lysis solution with proteinase K: immediately prior to use, add 0.5 mg proteinase K per ml of SDS lysis solution.

5.2 Single- and double-strand breaks (SSB and DSB)

DNA single- or double-strand breaks are assessed under fully deproteinizing conditions that eliminate protein adsorption to filters (1, 2) (see *Table 2*). The main characteristics of this type of elution are:

- Filters: non-protein adsorbent, polycarbonate (see Section 5.1)
- Lysis: SDS-lysis solution with proteinase K
- Wash: 0.02 M EDTA solution
- Elution: Pr_4NaOH-EDTA with SDS solution; pH 12.1 for SSB, pH 9.6 for DSB

The sensitivity of the SSB and DSB assays can be adjusted. The most commonly used assays are the high sensitivity methods ('long methods'), with slow pumping at 0.03–0.04 ml/min; 35–40 ml Pr_4NaOH elution solution; five fractions overnight at 3 h intervals. However, when SSB frequency > 1000 rad-equivalents or DSB frequency $> 10\,000$ rad-equivalents, the low sensitivity methods ('short methods') are used, i.e. pump flow rate: 0.12–0.16 ml/min; 12 ml Pr_4NaOH elution solution; six fractions at 5 or 10 min intervals (18, 22) (*Table 2*).

5.3 Protein-associated strand breaks (PASB) and protein-free breaks ('frank breaks', FB)

DNA strand breaks that are not protein-associated (e.g. protein-free, frank breaks) are measured under non-deproteinizing conditions that maximize protein adsorption to the filters. This filter elution assay is routinely done in combination with the single- and double-strand break elution assays (Section 5.2) to differentiate protein-associated (typically topoisomerase-mediated) and FB strand breaks (1, 2). The main characteristics of the PASB (and FB) elution are outlined in *Table 2*:

- Filters: Protein-adsorbent (PVC) (see Section 5.1)
- Lysis: LS-10 lysis solution
- Wash: 0.02 M EDTA solution
- Elution: Pr_4NaOH–EDTA, pH 12.1 elution solution without SDS

5.4 DNA–protein crosslinks (DPC)

Under non-deproteinizing conditions, DNA–protein crosslinks reduce the elution rate of single-strand breaks. Therefore, the treated cells are irradiated on ice immediately before the elution to introduce random single-strand breaks. The DNA–protein crosslinking elution assays utilize cells that had only been irradiated as reference (control) (1, 5) (see *Table 2*).

Random DNA single-strand breaks are generally produced by irradiating the cells with 30 Gy (approximately 1 SSB per 3×10^5 nucleotides).

Elution assays are then done as in Section 5.3 (non-deproteinizing conditions) (see *Table 2*).

5.5 Interstrand DNA crosslinks (ISC)

The presence of interstrand crosslinks reduces the elution rate of single-strand breaks that are produced by exposure to X- or gamma-rays. Therefore, to detect and quantitate interstrand crosslink frequencies, the cells are irradiated on ice immediately before the filter elution assays. DNA filter elution assays are done under deproteinizing conditions to eliminate the effect of potential DNA–protein crosslinks (1, 2) (*Table 2*).

The radiation dose is usually 3 Gy immediately before the elution. Elution assays are then done under fully deproteinizing conditions with pH 12.1 eluting solution containing SDS (as in Section 5.2).

5.6 Alkali-labile sites

Alkali-labile sites are gradually converted into single-strand breaks with increasing time and pH. Therefore, in SSB elution assays (see Section 5.2), alkali-labile sites are characterized by a non-linear elution rate that increases progressively with time (16, 20). The pH-sensitivity of the elution rate can be further tested by increasing the pH of the eluting solution (1, 2).

Experimental cells (control, calibration standard and drug-treated cells) are first divided in two sets of samples. Alkali-labile site assays are then done under fully deproteinizing conditions as in Section 5.2 with eluting solution at pH 12.1 for the first set of samples and at pH 12.6–12.8 for the second set of samples (pH of the eluting solution must be buffered using buffering solution (a) see Section 5.1)

5.7 Apoptosis-associated DNA fragmentation

The oligonucleosome-sized DNA fragmentation associated with apoptosis consists of small protein-free DNA double-strand breaks. Because of the short size of these DNA fragments, they elute rapidly within the lysis fractions. Because of this unique property, the total assay is much simpler and faster as it does not require further elution of the DNA. We refer to this assay as the 'DNA fragmentation' assay (9–12, 26).

Protocol 1. DNA fragmentation assay

1. Use non-deproteinizing conditions with protein-adsorbing filters and LS-10 lysis solution as in Section 5.3
2. Load cells (~ 0.5×10^6) directly on to the filter and wash rapidly with an additional 5 ml of balanced salt solution. Allow this solution to flow through by gravity.

Protocol 1. *Continued*

3. Add 5 ml of LS10 lysis solution and collect immediately.
4. Wash the lysis solution by adding 5 ml of 0.02 M EDTA solution pH 10.0. Remove the filter and process as in Section 4.4

Note: Other methods for detection and quantitation of apoptosis are described in Chapters 7 and 8.

5.8 Preparation of isolated nuclei for filter elution assays

DNA filter elution assays can also be done using isolated nuclei. This application is useful to bypass drug transport defects and to study subcellular localization of the DNA damaging pathways.

Protocol 2. Preparation of isolated nuclei

1. Wash cells twice in ice-cold nucleus buffer (see Section 5.1)
2. Incubate cells (1×10^7/ml) on ice with gentle rocking for 10 min in nucleus buffer containing 0.3% Triton X100.
3. Centrifuge (450 *g*, 10 min at 4°C) to pellet the nuclei. Supernatants contain the cytoplasm and plasma membrane debris.
4. Wash the nuclei pellets twice by centrifugation/resuspension in lysis buffer without Triton X100.

Isolated nuclei are then ready to be used for experiments

At the end of treatment, nuclei are processed in the various DNA filter elution assays as in the case of whole cells (see above).

5.9 Reconstituted cell-free system with filter elution assays

Endonuclease activities that are associated with apoptosis can be detected from a reconstituted cell-free system using the DNA filter binding assay (12, 26).

Protocol 3. Reconstituted cell-free system for apoptosis assays

1. Prepare nuclear and cytoplasmic fractions from control and apoptotic cells as in Section 5.8.
2. Centrifuge (12 000 *g*, 10 min at 4°C) supernatants obtained as in Section 5.8 to obtain soluble components of cytoplasmic fractions.

3. Incubate control nuclei with either control or drug-treated (apoptotic) cytoplasmic fractions at 30 °C for up to 30 min.
4. Load these extracts directly on to filters and carry out the DNA filter binding assay as in Section 5.7.

The reconstituted cell-free system in combination with DNA binding filter assay can be used to study the biochemical requirements for apoptosis. These assays can be used to study apoptosis-associated endonucleases and their regulatory factors following subsequent fractionation and purification of the cytoplasmic components (12, 26).

6. Computations

After isotope-counting, the data are calculated as the fraction of total DNA remaining on the filter for each fraction (see example in *Table 3*). This is done for both the ^{14}C- and the ^{3}H-labelled DNA. Usually the lysis fraction is excluded from computation and the fraction of DNA retained on the filter is therefore 1 for the lysis (+ wash) fraction (see *Table 3*).

A detailed analysis of the DNA lesion measurements has been published (1). Therefore, we will present the simplest computation methods that can be derived graphically. However, most laboratories where a large number of alkaline elution assays are performed have developed their own computer programs.

6.1 SSB (single-strand breaks)

Elution rates of random DNA single-strand breaks introduced by ionizing radiation yield first-order kinetics with respect to time (or internal standard)

Table 3. Example of data computation for a SSB elution

	d.p.m.	Fraction of total	Cumulative d.p.m.	Retention
Lysis + wash	200	0.042		1.000
Fraction no. 1	500	0.104	500	0.891
Fraction no. 2	600	0.125	1100	0.761
Fraction no. 3	800	0.167	1900	0.587
Fraction no. 4	400	0.083	2300	0.500
Fraction no. 5	200	0.042	2500	0.457
Line	50	0.010	2550	0.446
Wash	50	0.010	2600	0.435
Filter	2000	0.417	4600	0.000

The retention numbers (last column on the right) are used for plotting the fraction of DNA retained on the filter as a function of time (or fraction number). In the case where internal standard cells are used, ^{14}C-retention is plotted vs. ^{3}H-retention (see *Figure 3*).

and with respect to X- or gamma-ray dose (*Figure 3*). Based on this principle, the SSB frequency (expressed in rad-equivalents) can be calculated by the formula:

$$\text{SSB} = [\log(r_1/r_0)/\log(R_x/r_0)] \times X, \qquad [1]$$

where r_1 is the DNA retention for drug-treated cells, r_0 the DNA retention for untreated cells, R_x the DNA retention for irradiated cells that are used as calibrator, and X the irradiation dose (in rads). Any choice of the retention time or ^{3}H-retention on the x-axis of *Figure 3* is valid as far as the elution curves are first order, e.g. linear on a semi-log plot.

For the SSB 'long method' (high sensitivity assay), the irradiation dose for the calibrator and internal standard cells is usually 3 Gy (= 300 rads) and the formula becomes:

$$\text{SSB} = [\log(r_1/r_0)/\log(R_{300}/r_0)] \times 300 \qquad [2]$$

For the SSB 'short method' (low sensitivity assay), the irradiation dose for

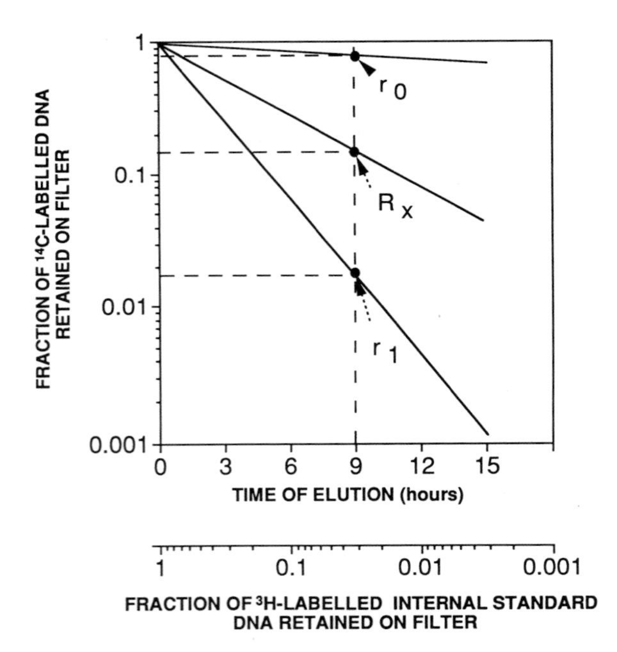

Figure 3. Theoretical example of semi-log plotting of DNA single-strand break (SSB) elution data. The data are plotted as a function of elution time on the x-axis. Alternatively, elution of internal standard cells can be used (lower x-axis). Three curves are shown. The upper curve is for untreated cells with r_0 as the retention at the chosen end point. The middle curve is for irradiated calibrator cells with R_x retention. The lower curve is for drug-treated cells with r_1 retention. SSB frequency for the drug-treated cells can be calculated according to Equations 1–3 (Section 6.1).

the calibrator and internal standard cells is usually 20 Gy (= 2000 rads) and the formula becomes:

$$SSB = [\log(r_1/r_0)/\log(R_{2000}/r_0)] \times 2000 \quad [3]$$

6.2 DSB (double-strand breaks)

Usually, DSB assays are performed without using internal standard cells in order to minimize the number of cells per filter (typically $< 5 \times 10^5$ cells/filter). Usually, elution curves are not first order, and elution rates tend to decrease with elution time, resulting in elution curves with a concave curvature. DNA retention at a given time point (usually 10 hours or 12 hours) is chosen to compute DSB frequency.

For a given cell line, it is recommended to generate a calibration curve by plotting the ^{14}C retention as a function of the irradiation dose. Usually, the curve fits better a linear–linear than a log–linear regression (*Figure 4*) (7, 13, 18), indicating that, by contrast to SSB, the elution rate of DSB is not first order. Using this calibration curve, it is possible to determine the DSB frequency graphically (*Figure 4*). Alternatively, once the equation of the calibration curve is generated, calculations can be performed directly. For instance in a previous publication we found that the DSB calibration curve for mouse L1210 nuclei was:

$$r = 0.903 - 0.088\ X, \quad [4]$$

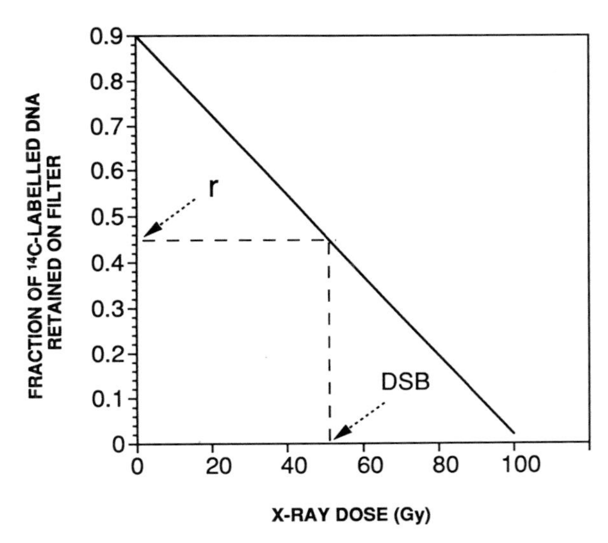

Figure 4. Theoretical regression curve of DNA retention as a function of X-ray dose. DSB in the drug-treated sample can be calculated by extrapolating the retention value of the drug-treated sample (*r*) into the X-irradiation curve. The results are usually expressed in rad-equivalents (100 rad-equivalents = 1 Gy-equivalent).

where r is the ^{14}C retention at 10 hours of elution and X the X-ray dose in rads. Therefore, if the ^{14}C retention for drug-treated cells is r_1, the DSB frequency (in DSB-rad-equivalent) is calculated as:

$$DSB = (0.903 - r_1)/0.088. \quad [5]$$

For drugs that produce both single- and double-strand breaks (such as topoisomerase II inhibitors), it is possible to calculate the ratio of the frequency of true single-strand breaks (s) (that are not part of a double-strand break) to the frequency of double-strand breaks (d) by the formula (7, 13):

$$s/d = (SSB/DSB) \times k - 2 \quad [6]$$

where SSB and DSB are the single- and double-strand break frequencies as calculated above, and k the ratio of true single-strand breaks per double-strand breaks in the case of X- or gamma-rays. This ratio has been estimated at around 23 (7, 13, 21). Therefore, if a drug produces solely double-strand breaks, s would be equal to 0, and the measured value for SSB/DSB would be 0.09.

6.3 DPC (DNA–protein crosslinks)

DNA retention in irradiated drug-treated cells is plotted against elution time (or fraction number) (*Figure 5*). The retention value that is used for calculation can either be derived by extrapolation of the last points (usually the last three points) of the elution curve to the y-axis (*Figure 5*, values indicated on the y-axis), or by taking a fixed time (or fraction) (*Figure 5*, values indicated at the 12 hour time).

Two basic equations have been derived for calculation of DPC frequencies depending on their distribution relative to DNA single-strand breaks (*Figure 1*). The random model is used when it is assumed that DPC and single-strand breaks are distributed randomly and independently of each other along the DNA (1, 2, 14, 17). The protein-associated model is used when it is assumed that DNA–protein crosslinks are at one terminus of the single-strand breaks as in the case of cleavable complexes produced by DNA topoisomerase inhibitors (22, 27). The formulas are as follows

For the random model:

$$DPC = [(1 - r)^{-1/2} - (1 - r_0)^{-1/2}] \times 3000 \quad [7]$$

For the bound to one terminus model:

$$DPC = [(1 - r)^{-1} - (1 - r_0)^{-1}] \times 3000 \quad [8]$$

where r is the retention for drug-treated cells, and r_0 the retention for cells that have not been drug-treated (control cells from another elution sample or internal standard cells for the same elution sample) (*Figure 5*).

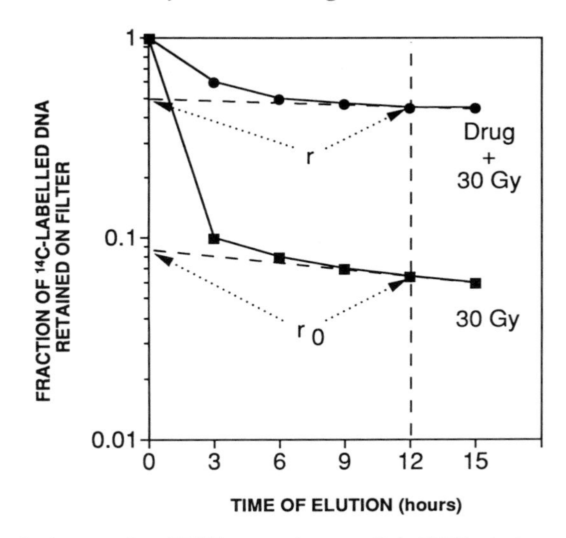

Figure 5. Theoretical example of DNA–protein crosslink (DPC) elution curves for control irradiated cells (30 Gy; lower curve) and drug-treated cells (drug + 30 Gy; upper curve). Retentions for control and drug-treated cells (r_0 and r, respectively) are obtained either as the retention at a given time of elution (arrows toward the right) or as the intercepts between the extrapolated elution curves (obtained from the last three elution points) and the *y*-axis (arrows toward the left). Control cells or the internal standard cells for the given drug-treated elution sample can be used as 'control'. DPC frequency for the drug-treated cells are calculated as described in Section 6.3 (Equation 7 or 8).

6.4 ISC (interstrand crosslinks)

Because elution curves of ISC are not linear but significantly curve, the elution slope does not give a unique measure for ISC. However, a good estimate of relative ISC frequency (in rad-equivalents) can be obtained by the formula:

$$\mathrm{ISC} = \{[(1 - R_0)/(1 - R_1)]^{1/2} - 1\} \times 300 \qquad [9]$$

where 300 represents the standard irradiation dose (3 Gy = 300 rads) used to introduce random SSB; R_0 represents the DNA retention of control cells treated only with the standard irradiation at a fixed elution endpoint; R_1 represents the DNA retention of experimental cells treated with an interstrand crosslinking drug and the standard dose of radiation at the same fixed elution endpoint (1, 2) (*Figure 6*).

6.5 Apoptosis-associated DNA fragmentation

DNA fragmentation is determined as the fraction of ^{14}C-labelled DNA in the lysis fraction + EDTA wash relatively to total intracellular DNA. Results

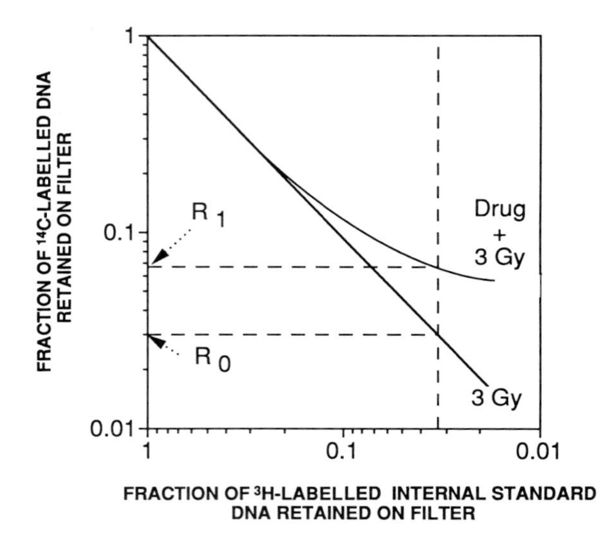

Figure 6. Theoretical example of DNA interstrand crosslink (ISC) elution curves for control irradiated cells (3 Gy; lower curve) and drug-treated cells (drug + 3 Gy; upper curve). Retentions for control (R_0) and drug-treated (R_1) samples are used to calculate the ISC frequency according to Equation 9 (see Section 6.4).

are expressed as the percentage of DNA fragmented in treated cells compared to DNA fragmented in control untreated cells (background) using the formula

$$[(F - F_0)/(1 - F_0)] \times 100 \quad [10]$$

where F and F_0 represent DNA fragmentation in treated and control cells, respectively (10–12, 26, 28). Experimental data are usually plotted as % of DNA fragmented versus time as shown in *Figure 7.*

7. Conclusions and perspectives

Except for apoptosis-associated DNA fragmentation that can be triggered by a variety of stimuli, all the other types of DNA lesions are specific and characteristic to subclasses of drug and chemical. Therefore, the DNA filter elution methodology is an important tool to study, characterize and classify the effects and mechanisms of action of new chemotherapeutic drugs and carcinogens. The DNA filter elution methodology is commonly used to monitor DNA damage formation and reversal following drug and chemical removal and to study the relationship between DNA damage and DNA repair, cell

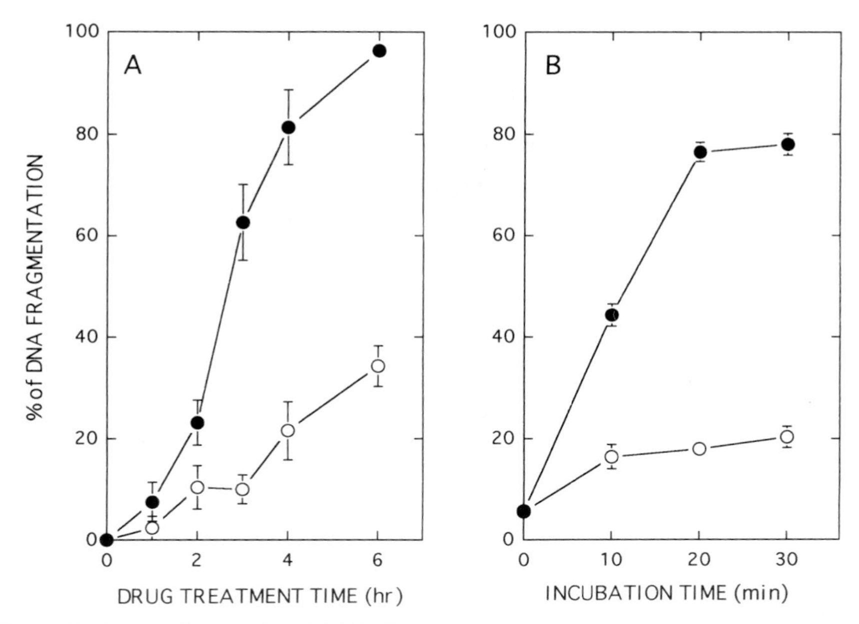

Figure 7. Apoptosis-associated DNA fragmentation measured by filter elution assay from cultured cells or from a reconstituted cell-free system. (A) HL-60 cells were treated at 0.1 μM (open circles) and 1 μM (closed circles) staurosporine concentrations. DNA fragmentation was determined by the DNA filter binding assay at specified times after drug removal. (B) DNA fragmentation induced in a reconstituted cell-free system by cytoplasmic fractions obtained from HL-60 cells treated with staurosporine (1 μM for 3 h). Cytoplasm from untreated HL-60 cells (open circles) or staurosporine-treated cells (closed circles) was incubated with nuclei isolated from untreated cells at 30°C for the indicated times. DNA fragmentation was determined by the DNA filter binding assay.

killing and tumour development. In cell growth and division control, DNA filter elution methodology is useful to study apoptosis kinetics and to investigate its modulation. In combination with a reconstituted cell-free system, the DNA binding filter assay provides a powerful tool to study the DNA fragmentation mechanisms and biochemical regulation of programmed cell death.

Acknowledgements

The authors wish to thank Dr Kurt W. Kohn, Chief of the Laboratory of Molecular Pharmacology, DTP, DCT, National Cancer Institute for his invaluable contribution to the development of the alkaline elution methodology, for his helpful advice and continuous support during the years that we have learned and developed the filter elution assays.

References

1. Kohn, K. W. (1991). *Pharmacol. Ther.*, **49**, 55.
2. Kohn, K. W., Ewig, R. A. G., Erickson, L. C., and Zwelling, L. A. (1981). In *DNA repair: a laboratory manual of research procedures* (ed. E. C. Friedberg and P. C. Hanawalt), pp. 379–401. Marcel Dekker, New York.
3. Kohn, K. W. (1986). *Basic Life Sci.*, **38**, 101.
4. Filipski, J. and Kohn, K. W. (1982). *Biochim. Biophys. Acta,* **698**, 280.
5. Filipski, J., Yin, J., and Kohn, K. W. (1983). *Biochim. Biophys. Acta*, **741**, 116.
6. Pommier, Y., Kerrigan, D., Schwartz, R., and Zwelling, L. A. (1982). *Biochem. Biophys. Res. Commun.*, **107**, 576.
7. Pommier, Y., Schwartz, R. E., Kohn, K. W., and Zwelling, L. A. (1984). *Biochemistry*, **23**, 3194.
8. Pommier, Y., Zwelling, L. A., Schwartz, R. E., Mattern, M. R., and Kohn, K. W. (1984). *Biochem. Pharmacol.*, **33**, 3909.
9. Bertrand, R., Sarang, M., Jenkin, J., Kerrigan, D., and Pommier, Y. (1991). *Cancer Res.*, **51**, 6280.
10. Bertrand, R., Solary, E., Jenkins, J., and Pommier, Y. (1993). *Exp. Cell Res.*, **207**, 388.
11. Solary, E., Bertrand, R., Jenkins, J., and Pommier, Y. (1993). *Exp. Cell Res.*, **203**, 495.
12. Solary, E., Bertrand, R., Kohn, K. W., and Pommier, Y. (1993). *Blood*, **81**, 1359.
13. Zwelling, L. A., Michaels, S., Erickson, L. C., Ungerleider, R. S., Nichols, M., and Kohn, K. W. (1981). *Biochemistry*, **20**, 6553.
14. Ross, W. E., Glaubiger, D., and Kohn, K. W. (1979). *Biochim. Biophys. Acta*, **562**, 41.
15. O'Connor, P. M. and Kohn, K. W. (1990). *Cancer Commun.*, **2**, 387.
16. Lafleur, M. V. M., Wolohuis, J., and Loman, H. (1981). *Int. J. Radiat. Biol.*, **39**, 113.
17. Ross, W. E., Glaubiger, D. L., and Kohn, K. W. (1978). *Biochim. Biophys. Acta*, **519**, 23.
18. Pommier, Y., Mattern, M. R., Schwartz, R. E., and Zwelling, L. A. (1984). *Biochemistry*, **23**, 2922.
19. D'Incalci, M., Covey, J. M., Zaharko, D. S., and Kohn, K. W. (1985). *Cancer Res.*, **45**, 3197.
20. Pommier, Y., Runger, T. M., Kerrigan, D., and Kraemer, K. H. (1991). *Mutat. Res.*, **254**, 185.
21. Blocher, D. (1982). *Int. J. Radiat. Biol.*, **42**, 317.
22. Pommier, Y. and Kohn, K. W. (1989). In *Developments in cancer chemotherapy* (ed. R. I. Gazer), pp. 175–96. CRC Press, Boca Raton, FL.
23. Szmigiero, L. and Studzian, K. (1988). *Anal. Biochem.*, **168**, 88.
24. Erickson, L. C., Osieka, R., Sharkey, N. A., and Kohn, K. W. (1980). *Anal. Biochem.*, **106**, 169.

25. Covey, J. M., Jaxel, C., Kohn, K. W., and Pommier, Y. (1989). *Cancer Res.*, **49**, 5016.
26. Bertrand, R., Solary, E., Kohn, K. W., and Pommier, Y. (1995). *Drug Develop. Res.*, **34**, 138.
27. Liu, L. F. (1989). *Annu. Rev. Biochem.*, **58**, 351.
28. Bertrand, R., Kerrigan, D., Sarang, M., and Pommier, Y. (1991). *Biochem. Pharmacol.*, **42**, 77.

7

Morphological and biochemical criteria of apoptosis

SUMANT RAMACHANDRA and GEORGE P. STUDZINSKI

1. Introduction

When cells receive mixed signals for growth they usually die. For instance, when the developmental programme requires cell division but external growth signals are lacking, or when a growth-related gene such as c-*myc* is highly expressed but the cellular environment has insufficient nutrient content, or a xenobiotic is present, the cell dies by a process termed apoptosis. Although there are differences in the phenomena observed during this sequence of events, depending on the cell type, and agent or circumstance which initiates the cell's demise, there are morphological and biochemical similarities which suggest that these are variants of the same biological process designed to control the size of cell populations.

It is important to distinguish apoptosis from the other major form of cell death, necrosis. First, at the tissue level, apoptosis produces little or no inflammation, since shrunken portions of the cell are engulfed by the neighbouring cells, especially macrophages, rather than being released into the extracellular fluid. In contrast, in necrosis, cellular contents are released into the extracellular fluid, and thus have an irritant effect on the nearby cells, causing inflammation. Second, there is the expectation that elucidation of the steps of the cellular mechanisms that lead to apoptosis may allow this form of cell death to be induced by cancer therapeutic agents. Third, the apoptotic mechanism of cell death is fundamental to the normal development of tissues and organisms. In contrast, cell death by necrosis does not have such significance.

The role of apoptosis in cell population control during development has suggested that there are inherent cellular programmes which lead the cell to self destruction. This has been confirmed in a number of instances, e.g. in a small nematode, *Caenorhabditis elegans* (*C. elegans*), where each individual cell can be recognized, it has been found that in the hermaphrodite form of the worm the same set of 113 cells is destined for programmed cell death during embryogenesis and another set of 18 cells later in life for a total of 131 cells (1). Also, inhibition of RNA or protein synthesis can, in many cases,

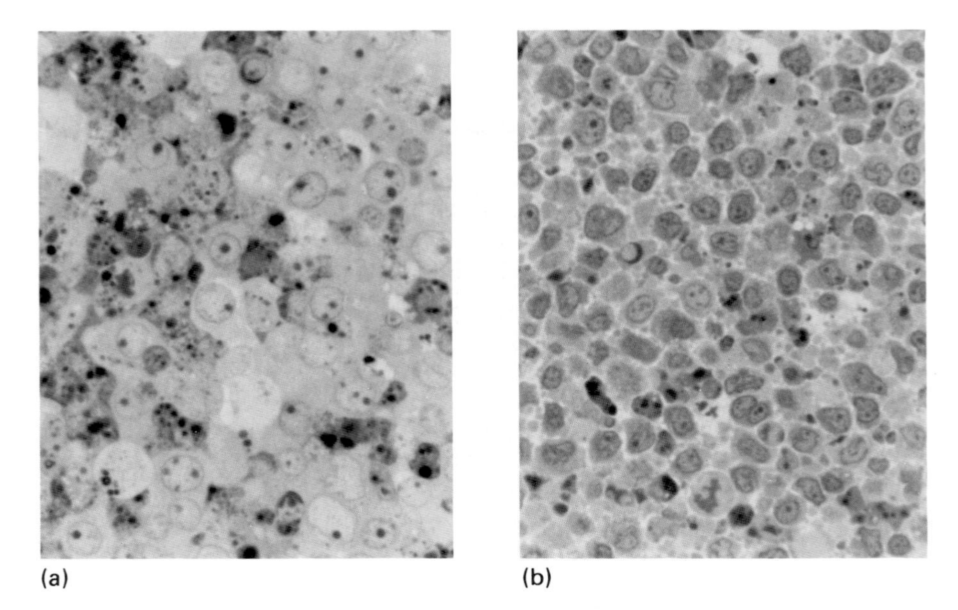

Figure 1. Illustration of light microscopic appearance of apoptotic cells and its modification by a differentiation-inducing agent. (a) HL60 cells were exposed to calcium ionophore A23187 (10 μM for 8 h), embedded in epon, and 1 μm sections were stained with Toluidine Blue. Note densely stained fragments of chromatin in nuclei and cytoplasm of most cells. (b) HL60 cells treated as in (a), but first exposed to 1,25-dihydroxyvitamin D_3 (10^{-8} M for 48 h) which protects HL60 cells against apoptosis (16). Note smaller (differentiated) cells, only a few of which show apoptotic nuclei.

abrogate cell death by apoptosis (2), although it usually accelerates necrosis. Thus, it appears that gene expression is necessary for cell death. Yet, there is another level of complexity, as, in some instances, inhibition of protein or RNA synthesis, or even expulsion of nuclei, does not prevent what otherwise appears to be programmed cell death (3). Such cells are thought to be 'primed' for apoptosis.

The original use of the term 'apoptosis' was primarily descriptive of the cellular morphology of dying cells (4). Although some authors blur the precision of the term (3), it is still tenable to define apoptosis as cell death that differs from necrosis on a morphological basis, observable by light or by electron microscopy. The key features originally described included shrinkage and blebbing of the cytoplasm, preservation of structure of cellular organelles including the mitochondria, and condensation and margination of chromatin, though not all of these are seen in all cell types (*Figure 1*). It is generally assumed that these morphological changes result from a developmental programme for cell death that can be triggered by a deprivation of a growth factor, or by addition of a xenobiotic compound such as a cancer therapeutic drug. The morphological criteria are still the most important

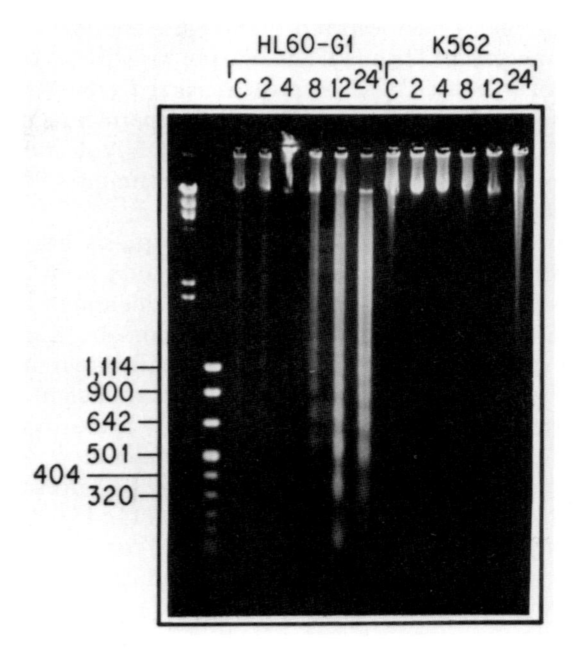

Figure 2. DNA ladder formation in HL60-G1 cells (a subclone of human leukaemia HL60 cells), but not in K562 human leukaemia cells, following exposure to doxorubicin (5 μM for HL60 cells and 10 μM for K562 cells, both for 24 h). DNA was extracted and run on 2% agarose gels and stained with ethidium bromide. DNA ladders indicative of internucleosomal DNA fragmentation became apparent after 8 h, coincident with morphological appearances of apoptosis (not shown). Microscopic examination of doxorubicin-treated K562 cells showed that these cells became necrotic (not shown). Reprinted from ref. 9 with permission from Wiley-Liss, Inc., a division of John Wiley & Sons, Inc.

when complex cell populations, such as tissues, are examined, and overall cell shrinkage and nuclear condensation are the easiest to recognize.

In pure cell populations, biochemical changes in chromatin and DNA degradation provide useful and often quantifiable means of detecting apoptosis. It is often forgotten, however, that random DNA degradation is not a specific test for apoptosis but simply demonstrates cell death. Although detection of DNA degradation may be useful as an adjunct method of quantitation, occurrence of apoptosis has to be shown by morphological or by more specific biochemical methods.

The classical biochemical method for demonstrating apoptosis is the presence of oligonucleosome-sized fragments of DNA, which, when run on agarose gels, produce 'ladders', as illustrated in *Figure 2* (5, 6). It has recently been shown that an earlier endonucleolytic cleavage of chromatin produces DNA fragments from 300 kb down to 50 kb in size (7, 8). Also, the observation that mitochondrial DNA is intact in early stages of apoptosis, provides a basis for a new method which can detect and quantify apoptosis (9). These methods do prove that apoptosis has occurred, although sometimes a lag period of several hours is necessary for the signs of apoptosis to become detectable. Demonstration of increased nuclear protease activity may also add to this evidence (10). In contrast, increased membrane permeability which allows dye entry, e.g. of Trypan Blue, demonstration of DNA strand breaks, thymidine release, and dozens of other purported techniques for detection of apoptosis are relatively non-specific.

The programmed nature of cell death by apoptosis has suggested that expression of some genes is specifically associated with apoptosis. For instance, in *C. elegans*, several genes have been identified that function in different steps of the genetic pathway of programmed cell death (11). One of the essential genes for initiation of the cell death programme, *ced*-3 was found to encode a protein similar to the mammalian interleukin-1β-converting enzyme (ICE), and to the product of a murine gene, *nedd*-2, expressed in the embryonic brain (12). In mammalian prostatic cells a candidate for such a gene is *TRPM*-2 (testosterone-repressed prostatic message-2) (13), and in T-lymphocytes the *fas*/*APO*-1 gene (14, 15). Other candidate genes for an apoptosis-related role include *p53*, c-*fos*, transforming growth factor β1, c-*myc*, *ras*, ornithine decarboxylase and calmodulin (3). Whether a general role in apoptosis for any of these genes can be established awaits further studies. At this time only ICE, *fas*, *bcl*-2 and associated genes, e.g., *bcl*-x, *bax*, *mcl*-1, and to a lesser degree c-*myc*, appear to be directly involved in the regulation of human cell survival and death.

2. Distinction of apoptosis from other forms of cell death

Current literature contains many examples of a loose use of the term 'apoptosis'. There are several forms of cell death, and in all of them nuclear DNA becomes at some point degraded. As mentioned above, demonstration of DNA damage or release of products of DNA degradation is by itself insufficient to justify the description of the phenomenon as apoptosis. The distinction is more than a semantic debate, since the concept behind the term 'apoptosis' is the existence of an inherent cellular programme, somewhat similar to the programmes which drive cell differentiation, whereas 'necrosis' results entirely from circumstances outside the cell. Other forms of cell death, e.g. mitotic cell death (see Chapter 8), are insufficiently characterized to be considered at this time as biologically distinct entities.

The criteria most useful for distinguishing apoptosis from necrosis are listed in *Tables 1* and *2*. Importantly, there are also similarities between apoptosis and necrosis, and in view of these, and a frequent overlap in characteristic features, conclusive evidence of the occurrence of apoptosis should demonstrate more than one morphological or biochemical criterion of apoptosis.

Table 1. Morphological differences and similarities between apoptosis and necrosis

	Differences		Similarities or confounding variables
	Apoptosis	**Necrosis**	
1. Nuclei	Pyknosis and karyorrhexis (dense condensation of chromatin)	Karyolysis, preceded by irregular chromatin clumping	Damage occurs in both
2. Cytoplasmic organelles	Intact	Disrupted	2^{ry} damage in apoptosis
3. Cell membrane	Apoptotic bodies blebbing	Blebbing and loss of integrity	Changes seen in both
4. Cell volume	Cells shrink	Cells swell	There may be no detectable changes
5. In tissues	Single cells affected	Groups of cells affected	In epithelia, superficial cells are apoptotic and in groups
6. Tissue response	None	Inflammation	—

3. Morphological changes in apoptosis

Morphological changes in apoptosis have been described in detail in early publications describing apoptosis (e.g. 4, 16). Arends *et al.* (17) divided the morphological changes of apoptosis into three phases. In the first phase, there is condensation of chromatin into crescent-shaped caps at the nuclear periphery, disintegration of the nucleoli, and reduction in nuclear size. Also observed are shrinkage of total cell volume (hence the original name of 'shrinkage necrosis' given to the cells undergoing apoptosis), an increase in cell density, compaction of some cytoplasmic organelles, and dilatation of the endoplasmic reticulum. Through these early changes, the mitochondria remain morphologically normal. In the second phase (which may overlap with the first phase), there is budding (blebbing) and constriction of both the nucleus and the cytoplasm into multiple, small, membrane-bound apoptotic bodies. *In vivo*, these apoptotic bodies may be shed from epithelial surfaces or phagocytosed by neighbouring cells, usually macrophages. In the third phase, there is progressive degeneration of residual nuclear and cytoplasmic structures with characteristics which resemble those of necrosis, and therefore referred to as 'secondary necrosis' (17). The morphological hallmark of apoptosis is the condensation of the nucleus, whereas other organelles are relatively well maintained (4).

Phase contrast microscopy of cultured cells undergoing apoptosis shows that the visible evidence of the apoptotic process is of sudden onset, and that

Table 2. Biochemical differences and similarities between apoptosis and necrosis

	Differences		Similarities or confounding variables
	Apoptosis	**Necrosis**	
1. Nuclear DNA damage	Nucleosomal and/or 50–300 kb fragments → ladders on gels	Random → smears on gels	Takes place in both but more in apoptosis
2. Nuclear gene expression	Usually needed	Not needed	Not needed in cells 'primed' for apoptosis
3. Mitochondrial DNA damage	Spared	Occurs early	—
4. Enzyme activation			
(a) DNases	Necessary	Not necessary	Lysosomal DNase and proteases are activated in necrosis
(b) Proteases	Necessary		
(c) Transglutaminase	Frequent		
5. Membrane function	Intact	Loss of function	—
6. Cell internal milieu			
(a) pH	Slightly acidic (pH 6.4)	Acidic	Both acidic
(b) Ca^{2+}	Often increases	Always increases	Seen in both
(c) Na^+/K^+ pump	May be intact	Defective	

its early stages appear to occur swiftly. The early microscopic changes of rounding-up, rapid cellular shrinkage and violent pulsation, blebbing of the membrane, and formation of apoptotic bodies is often completed within a few minutes. The average period for which an apoptotic body formed in a tissue remains visible with the light microscope is between 12 and 18 h (16).

3.1 Identification by light microscopy of cells undergoing apoptosis

Light microscopy is the simplest way in which apoptosis can be recognized in cell suspensions or tissue sections. Identification of the apoptotic cells by this method can then be confirmed by other techniques such as electron microscopy, and histochemical or biochemical studies.

3.1.1 Cell suspensions

In preparing cell suspensions for light microscopic staining, several points have to be taken into consideration. The cell suspension should be free of debris or clumps of cells. Washing the cells in an isotonic physiological solution, such as phosphate-buffered saline (PBS, pH 7.2) usually removes much of the protein debris. If clumps are seen, one can use gentle mechanical disruption such as pipetting the clumps up and down to get the aggregated cells back into suspension. If that fails, the clumped cells can be removed from the suspension by letting the aggregate(s) settle then pipetting the suspension into another tube.

Cytospin preparations are, in general, most suitable for the demonstration of apoptotic morphology of cell suspensions. An important consideration is the concentration and the volume of the cells added to a cytospin cup. If the cells are too concentrated, microscopic study will be difficult since the cells may overlap each other. On the other hand, if the cell concentration is too low, the cells will be widely dispersed, and too few cells will be available for an accurate examination. Generally 0.5 ml of a $1–5 \times 10^5$ cells/ml suspension is appropriate. Addition of fetal calf serum (FCS) helps to stabilize cell membranes. In enumerating the proportion of apoptotic cells in a cytospin preparation all possible microscopic fields should be examined because the apoptotic cells tend to be distributed unequally, usually towards the periphery of the area with sedimented cells. Staining the cytospin preparation with Mayer's 0.1% haematoxylin is very simple and totally adequate for the purpose (18). However, virtually any dye that combines with DNA can be used, e.g. DNA-binding fluorophore Hoechst dyes and propidium iodide are also frequently used for demonstration of apoptotic cells, but require more elaborate equipment, i.e., a fluorescent microscope, and offer no special advantage over haematoxylin stain, although they are useful for flow cytometric analysis of apoptosis as described in Chapter 8.

The haematoxylin-stained slides can be counterstained with eosin Y, but this is usually not necessary. If another stain is preferred, the method used is the same as in *Protocol 1* except that this stain is used in place of the Mayer's 0.1% haematoxylin. Intensity of a water-soluble stain such as Giemsa can be increased by increasing the time the slide is exposed to the stain, and if the cells are over stained additional washing steps with water will reduce the intensity.

Protocol 1. Preparation of cytospin slides and haematoxylin staining

Note: Except when otherwise indicated, all reagents listed in all protocols in this Chapter are available from Sigma Chemical Co.

Equipment and reagents

- Cytospin and cytospin cup (Shandon)
- Pre-cleaned frosted-end slide and coverslip (Fisher)
- Phosphate-buffered saline (PBS) tablets, make 1× solution, pH 7.2, supplement to 5% FCS
- Ethanol (absolute) (EM Science)
- Mayer's 0.1% haematoxylin (all reagents from Fisher): dissolve 1 g haematoxylin indicator in 1 litre distilled water, heat solutions to boiling, and boil for 5 min. Remove from heat and add 0.2 g sodium iodate when the boiling stops. Let it sit for 10 min then add 50 g aluminium potassium sulphate, 1 g citric acid, 50 g chloral hydrate in this order, making sure each reagent is dissolved completely before adding the next. The solution can be used for 2–3 months

A. *Cell preparation*

1. Prepare cell suspension from culture or fresh preparation.
2. Pellet the cells by centrifugation (1200 *g* for 5 min at 4°C).
3. Resuspend in cold PBS to 5×10^5 cells/ml.
4. Add 0.5 ml (2.5×10^5 cells/slide) of suspension to prepared cytospin cup and slide setup.
5. Cytocentrifuge (6 min at 1200 *g*).

B. *Staining*

1. Fix the cytospin preparation in absolute alcohol: 10 dips.
2. Stain the slide in 0.1% haematoxylin for 5 min.
3. Wash briefly in distilled water, then rinse gently in tap water.
4. Coverslip the cytospin and fix slip in place.
5. Cytospin slides are ready for microscopic examination.

Under some circumstances an investigator may prefer a fluorescence-based method to detect nuclear DNA condensation characteristic of apoptosis. A protocol is therefore provided for staining with propidium iodide (PI) and examination with a fluorescent light microscope. In some cell lines staining with Acridine Orange gives excellent results (19), as discussed in Chapter 8.

Protocol 2. Identification of apoptotic nuclei by fluorescence microscopy

Equipment and reagents

- Fluorescent light microscope (Carl Zeiss)
- Pre-cleaned slides and coverslips (Fisher)
- Ethanol (EM Science) 70%, cold
- PBS, pH 7.2
- RNase, Type 1-A 1 mg/ml in PBS
- PI, 100 μg/ml in PBS

Method

1. Pellet the cell suspension (1×10^6 cells) by centrifuging at 200 *g* for 5 min.
2. Decant the supernatant carefully, and resuspend the pellet gently in the fluid remaining above the pellet.
3. Fix the resuspended pellet with cold 70% ethanol to approximately the original cell concentration, and keep at 4°C for 60 min.
4. Centrifuge at 200 *g* for 5 min, and wash with 1 ml PBS.
5. Resuspend in 0.5 ml PBS and add an equal volume of RNase.
6. Mix gently, then add 1 ml of PI solution.
7. Incubate the mixed cells in the dark at room temperature for 15 min, then keep at 4°C in the dark until ready for examination.
8. Examine stained samples on a coverslipped slide with a fluorescent microscope. 1250× magnification gives good resolution.

3.1.2 Tissue sections

Examination of tissue sections for the presence of apoptotic cells allows the investigator to study the process of cell death with the native tissue architecture intact. In many instances, cells at various stages of apoptosis can be recognized in relation to the surrounding cells. This allows for the recognition of the interaction of the apoptotic cell with the other elements in its environment. In addition, tissue sections help the investigator localize where the process of apoptosis is taking place (e.g., cortex vs. medulla of the lymph nodes).

Tissue preparation is more complicated than that for cell suspensions. Fresh solvents and solutions should be used and instructions should be followed closely for good staining. This procedure can also be used for paraffin or epon-embedded cell pellets. Care must be taken to ensure that the tissue sections are not extensively damaged while handling, as this disrupts cellular architecture.

Histological techniques are well described in ref. 20.

Protocol 3. Preparation of tissue sections for staining

Equipment and reagents

- LKB Nova microtome (Pharmacia)
- Slides pre-coated with 0.01% poly-L-lysine: dissolve poly-L-lysine (300 000 mol. wt) to 0.01% in sterile distilled deionized water. Treat slides (Fisher) with the 0.01% solution and allow to air dry
- Formalin solution, neutral buffered 10%
- Paraffin (Fisher)
- Xylene (Fisher)
- Ethanol (EM Science), 96%, 90%, 80%

A. *Tissue preparation*

1. Fix tissue samples in buffered formalin.
2. Embed the sample in paraffin.
3. Cut 4–6 μm sections of the paraffin-embedded tissue.
4. Adhere sections to the poly-L-lysine-treated slides.
5. Deparaffinize the sections by heating the sections for 10 min at 70°C or 30 min at 60°C.
6. Hydrate the sections by treating the slides in the following manner:
   (a) Soak them twice in a xylene bath for 5 min each time.
   (b) Then, soak in each of the following for 3 min each:
      i. 96% ethanol
      ii. 90% ethanol
      iii. 80% ethanol
      iv. Deionized distilled water (DDW)

B. *Staining*

1. Follow *Staining* steps in *Protocol 1* (from step 2).

3.2 Identification by electron microscopy of cells undergoing apoptosis

The electron microscope is an important tool in the study of the complex morphological changes that occur in apoptosis. Early changes such as the formation of chromatin crescents and cytoplasmic condensation can easily be noted by transmission electron microscopy (TEM). Fine details of the process, such as the disintegration of the nucleolus, or the deep convolutions of the nuclear membrane, can also be observed. TEM gives a bird's eye view of the processes occurring within a cell at the macromolecular level. Scanning electron microscopy (SEM) gives a 3-dimensional view of the apoptotic process, and displays the dramatic protruberances (blebbing) of the cell and nuclear membrane in the formation of apoptotic bodies. Electron microscopic studies have revealed the multi-step nature of the process of apoptosis and have been extensively described (e.g. 4, 16, 17).

The procedures for the preparation of cells or tissue sections for electron microscopy require advanced training and are described in specialized texts.

4. Histochemical detection of apoptosis

Demonstration of the products of chemical reactions at tissue level for detection of cell damage has an advantage in that the investigator simultaneously obtains morphological data and information on the localization of the reaction products. The fragmentation of nuclear DNA which occurs in apoptosis can be detected by labelling the newly formed free ends of DNA, followed by the examination of the label under a light microscope. A technique has been described for the histochemical detection of chromosomal breaks based on a TdT-mediated dUTP biotin nick end-labelling (TUNEL) of fragmented nuclear DNA *in situ* (21). A variant of this method, *in situ* end-labelling (ISEL) of the fragmented DNA, differs from TUNEL in its use of DNA polymerase in place of the terminal deoxynucleotidyl transferase (TdT) (22). Because of this similarity, only the TUNEL method is described here. It can be used on tissue sections or on cell suspensions. In addition, a commercial kit has been developed by Oncor, called Apoptag, that utilizes this technique.

Protocol 4. Labelling of nuclear DNA fragments in fixed tissue sections by the TUNEL method

Equipment and reagents

- LKB Nova microtome (Pharmacia)
- Slides pre-coated with 0.01% poly-L-lysine: dissolve poly-L-lysine (300 000 mol. wt) to 0.01% in sterile distilled deionized water. Treat slides with the 0.01% solution and allow to air dry
- Formalin solution, neutral buffered 10%
- Paraffin (Fisher)
- Xylene (Fisher)
- Ethanol, 96%, 90%, 80%
- Proteinase K (PK), 20 μg/ml
- H_2O_2, 2%
- Terminal deoxynucleotidyl transferase (TdT) (Gibco BRL)
- TdT buffer: aqueous solution with 30 mM Trizma base (pH 7.2), 140 mM sodium cacodylate, 1 mM cobalt chloride
- Biotinylated dUTP
- Terminating buffer: aqueous solution with 300 mM sodium chloride and 30 mM sodium citrate
- Bovine serum albumin (BSA), 2% aqueous solution
- Peroxidase conjugated streptavidin, 1:10 dilution in water
- DAB/H_2O_2 solution: dissolve 0.2 mg/ml diaminobenzidine (DAB) tetrachloride in 50 mM Tris–HCl with 0.005% H_2O_2
- DN buffer: aqueous solution of 30 mM Trizma base, pH 7.2, 140 mM potassium cacodylate, 4 mM $MgCl_2$, 0.1 mM DTT
- DNase I (Gibco BRL)

A. *Tissue preparation*

1. Fix tissue samples in buffered formalin.
2. Embed the sample in paraffin.
3. Cut 4–6 μm sections of the paraffin-embedded tissue.

Protocol 4. *Continued*

4. Adhere sections to the poly-L-lysine treated slides.
5. Deparaffinize the sections by heating for 10 min at 70°C or 30 min at 60°C.
6. Hydrate the sections by treating the slides in the following manner:
   (a) Soak the slides twice in a xylene bath for 5 min each time.
   (b) Then, soak in each of the following for 3 min each:
      i. 96% ethanol
      ii. 90% ethanol
      iii. 80% ethanol
      iv. Deionized distilled water (DDW)

B. *DNA nick end-labelling of tissue sections*

1. Treat the tissue sections with PK for 15 min at room temperature (RT) to strip the proteins from the nuclei.
2. Wash the slides four times in DDW for 2 min each.
3. Inactivate the endogenous peroxidase by covering the sections with 2% H_2O_2 for 5 min at RT.
4. Rinse the sections with DDW and immerse the section in TdT buffer.
5. Add TdT (0.3 U/ml, final concentration) and biotinylated dUTP in TdT buffer (0.4 nmol/ml, final) to the buffer immersing the section.
6. Incubate the slides in a humid atmosphere at 37°C for 1 h.
7. Stop the reaction by transferring the slides to the terminating buffer for 15 min at RT.
8. Rinse the slides with DDW and cover with 2% aqueous solution of BSA for 10 min at RT to block non-specific binding.
9. Rinse the slides in DDW and immerse in PBS for 5 min.
10. Incubate the sections with peroxidase-conjugated streptavidin for 30 min at RT.
11. Rinse the slides in DDW and immerse in PBS for 5 min.
12. Stain the slides with the DAB/H_2O_2 solution for 30 min at 37°C.
13. Sections can be counterstained with another dye such as methyl green.
14. For a positive control specimen:
    (a) Pretreat tissue sections with DN buffer.
    (b) Add DNase I (1 μg/ml, dissolved in DN buffer) to cover the section for 10 min at RT.
    (c) Wash slides extensively with DDW and continue processing through the DNA nick end-labelling step (step 1).

15. For negative control specimens, treat the section like the sample section but use a buffer lacking TdT (step 5).
16. TUNEL score is recorded as the percentage of positively stained nuclei to total nuclei. Assess this by counting at least 1000 nuclei in high power (e.g., 400×) representative fields.

Protocol 5. Labelling of nuclear DNA fragmentation in cell suspensions by the TUNEL method

Equipment and reagents

- Same as for *Protocol 4*

A. *Preparation of cell suspension for cytology*

1. Centrifuge cell suspension for 4 min, 400 *g*, at RT.
2. Discard supernatant.
3. Resuspend the pellet in the remaining medium.
4. Add 4% buffered formaldehyde to the suspension and keep at RT.
5. Add a drop of the cell suspension on to a 0.01% poly-L-lysine coated slide and allow to air dry.

B. *DNA nick end-labelling of cell suspension preparations*

1. Treat as for tissue preparations (from step 1) except for the following:
   (a) No deparaffinization is needed.
   (b) Omit the incubation with PK.
2. Perform the analysis as in *Protocol 4.*

5. Biochemical detection of specific DNA damage for demonstration of apoptosis

It is clear that an endonucleolytic pathway is activated in apoptosis and results in the cleavage of the cell genome, first into large DNA fragments that vary from 300 kb to 50 kb in size, then into 180 bp multiples (7, 8). This has been referred to in Chapter 6 and illustrated in *Figure 1* of that chapter. Together with morphological changes characteristic of the process of apoptosis, the 180 bp periodicity of the DNA fragments visible on agarose gels stained for DNA is a tell-tale sign of this mode of cell death. It is believed that the DNA ladder formation by low-molecular-weight DNA fragments results from the nucleolytic cuts in linker DNA between nucleosomes within chromatin; the DNA derived from one nucleosome and the linker region is about 180 bp long. Cuts which lead to oligosome-sized fragments result in multiples of 180 bp. The long-term stability of the extracted DNA also adds favour to these procedures because the DNA samples can be analysed weeks after extraction. However, nucleosomal ladders are not essential for demonstration of apoptosis, since the process can occur without this type of DNA fragmentation (e.g. 23, 24).

The procedure for the analysis of nucleosomal DNA fragments is divided here into protocols for DNA isolation and a protocol for agarose gel electrophoresis, because several methods can be used which isolate DNA with varying proportions of high- to low-molecular-weight components. Each approach has special advantages. If high-molecular-weight DNA is included, the extent of apoptosis in different cell or tissue samples can be estimated by the relative proportion of DNA in the upper part of gel, as illustrated in *Figure 2*. This procedure is described in *Protocol 6* and can also be used to isolate and examine DNA for other purposes, e.g. the determination of mitochondrial DNA to nuclear DNA ratio described in *Protocol 9*. Methods which select for low-molecular-weight DNA, such as variants of the Hirt lysis procedure (e.g. ref. 25), increase sensitivity of detection of apoptosis. A recently described procedure for selective extraction of degraded DNA from apoptosis cells (26) is also presented here as *Protocol 7*.

Protocol 6. Isolation of DNA of high and low molecular weight

Equipment and reagents

- Beckman J-6 centrifuge or equivalent
- Digestion buffer: aqueous solution of 100 mM NaCl, 10 mM Tris–HCl (pH 8.0), 25 mM EDTA (pH 8.0), 0.5% SDS, 0.2 mg/ml proteinase K (PK)
- Phenol/chloroform/isoamyl alcohol (25:24:1) (Ameresco)
- Chloroform/isoamyl alcohol (24:1) (Ameresco)
- Ammonium acetate, 7.5 M
- Ethanol (EM Science), 100%, 70%
- TE buffer (pH 8.0): aqueous solution of 10 mM Tris–HCl, 5 mM EDTA

Method

1. Lyse cells with the digestion buffer.
2. Incubate lysed cells at 50°C for 12 h.
3. Extract lysates once with phenol/chloroform/isoamyl alcohol, and twice with chloroform/isoamyl alcohol.
4. Precipitate the DNA by adding ammonium acetate to 2.5 M, mix well, then add 2 volumes of 100% ethanol.
5. Incubate at 4°C for at least 2 h.
6. Pellet the DNA by centrifugation at 4000 r.p.m. (12 000 × *g*) for 30 min at RT.

7. Wash pellets with 70% ethanol and repeat centrifugation step.
8. Aspirate supernatant and air dry the pellet.
9. Dissolve the pellet in TE buffer.
10. Take aliquot for spectrophotometric readings at 260 nm and 280 nm.
11. The DNA extract obtained as described here can be used in *Protocol 8*, or in *Protocol 10*.

Protocol 7. Selective procedure for extraction of low-molecular-weight DNA from apoptotic cells

Equipment and reagents

- SpeedVac concentrator (Savant) or equivalent
- Ethanol 70%
- Hank's buffered salt solution (HBSS)
- Phosphate–citrate (PC) buffer: mix 192 parts of 0.2 M Na_2HPO_4 and 8 parts of 0.1 M citric acid and adjust to pH 7.8
- Nonidet NP-40, 0.25% solution in distilled water
- RNase A, 1 mg/ml solution in distilled water
- Proteinase K, 1 mg/ml solution in distilled water
- Loading buffer: aqueous solution with 0.25% bromophenol blue, 0.25% xylene cyanol FF and 30% glycerol

Method

1. Transfer, with a Pasteur pipette, 1 ml of cell suspension (2×10^6 cells/ml) in HBSS into tubes containing 10 ml of 70% ethanol, on ice.
2. Store the cells in the ethanol fixative for 24 h at room temperature (RT) or at −20°C.
3. Centrifuge the cells at 800 *g* for 5 min and remove every trace of the ethanol.
4. Resuspend the cell pellets in 40 μl of PC buffer at RT and transfer the suspension to 0.5 ml Eppendorf tubes.
5. Incubate for at least 30 min at RT.
6. Centrifuge the samples at 1000 *g* for 5 min.
7. Transfer the supernatants into a new tube and concentrate using a SpeedVac or equivalent for 15 min.
8. Add 3 μl of the Nonidet NP-40 solution and then add 3 μl of the RNase A solution.
9. Incubate for 30 min at 37°C.
10. Add 3 μl of the proteinase K solution and incubate again for 30 min at 37°C.
11. The DNA extract obtained as described here is mixed with 12 μl of loading buffer and can be used in *Protocol 8* (step 2).

Protocol 8. Analysis of nucleosomal DNA fragments by agarose gel electrophoresis

Equipment and reagents

- Electrophoresis apparatus (BioRad)
- UV transilluminator (Fisher)
- Polaroid photography equipment (Fisher) or a gel documentation system e.g., The Imager (Appligene)
- DNase-free RNase (Boehringer-Mannheim)
- Agarose powder, molecular biology grade
- TBE buffer, 10 × (Gibco BRL)
- Agarose gel 2%: dissolve 2 g of agarose in 100 ml 0.5 × TBE (Gibco BRL 10×) buffer by heating to a boil and stirring. Let the gel cool to about 50°C and pour to the desired gel mould to solidify. For 0.8% gels, use 0.8 g of agarose in 100 ml of 0.5 × TBE.
- Ethidium bromide, 10 mg/ml in distilled water (stock solution)

Method

1. Take 10 μg of DNA sample isolated in *Protocol 6* and incubate it with 0.1 U of DNase-free RNase for 1 h at 37°C.
2. Load the samples into the wells of a 2% agarose gel. *Note*: When low-molecular-weight DNA predominates in the sample to be analysed, e.g. when obtained by the modified Hirt protocol (25), or by *Protocol 7*, use 0.8% agarose gel.
3. Electrophorese in 0.5 × TBE buffer at 4 V/cm for 7 h.
4. Stain the gel with ethidium bromide (0.5 μg/ml in distilled water) for 30 min.
5. Destain the gel for 1–2 h in several changes of water.
6. Place the gel on a UV illuminator box and observe the samples for the 180 bp multiples (*Figure 2*).
7. Image can be recorded by one of the following methods:
   (a) Photograph with a Polaroid camera with a UV adapter.
   (b) Save the image on a gel documentation system. The advantage of this method is that image can be saved on disk and analysed later.
8. The presence of bands with 180 bp periodicity ('DNA ladder') indicates that at least some of the cells in the population under study died by apoptosis.

It has been recently reported that before the characteristic 180 bp multiples DNA ladders can be detected, very high-molecular-weight DNA fragments can be demonstrated by pulsed field gel electrophoresis of apoptotic cells (7, 27). The initial cleavage into fragments approximately 700, 300 and 50 kb is followed by an endonuclease cleavage into fragments of 180–250 bp multiples (8). The 300 kb fragments are considered to be generated by activation of an endogenous endonuclease or topoisomerase II molecules located at specific sites in chromatin, perhaps the nuclear matrix, whereas the 50 kb fragments and the accompanying DNA ladders are generated by endonucleolytic activity which appears to increase during apoptosis. The various sizes of the very large fragments have been suggested to be related to the higher order of folding of chromatin into chromosomes (28).

It is of interest to note that some studies have shown that the 300 and/or 50 kb DNA fragments can be demonstrated in morphologically characterized apoptotic cells, whereas the 180 bp multiple DNA fragments cannot (8, 27). Thus, the infrequent nuclease cuts of DNA appear to represent the earlier, and perhaps more fundamental, aspect of apoptosis.

Protocol 9. Analysis of high-molecular-weight DNA fragments by pulsed field electroporesis

Equipment and reagents

- Hypodermic needle (Becton-Dickinson)
- Glass moulds (BioRad)
- Multi-purpose rotator (Fisher)
- Horizontal gel electrophoresis tank (BioRad)
- Model 200/20 power supply (BioRad)
- Pulsewave 760 field inverting accessory (BioRad)
- Nuclear buffer: aqueous solution of 0.15 M NaCl, 2 mM KH_2PO_4/KOH (pH 6.4), 1 mM EGTA, 5 mM $MgCl_2$
- Proteinase K, 2 mg/ml
- Low melting point (LMP) agarose: 1.5% LMP agarose is prepared in nuclear buffer
- Lysis buffer: aqueous solution of 10 mM NaCl, 10 mM Tris-HCl (pH 9.5), 25 mM EDTA, 10% *N*-lauroyl sarcosine
- Storage buffer: aqueous solution of 10 mM Tris-HCl (pH 8.0), 1 mM EDTA
- High-molecular weight DNA size markers (BioRad)
- TAE buffer: aqueous solution of 40 M Tris-acetate (pH 8.5), 4 mM EDTA
- Agarose, 1.5% prepared in 0.25 × TAE
- Ethidium bromide, 10 mg/ml in distilled water (stock solution)

A. *Preparation of agarose plugs from cells*

1. Resuspend 1×10^7 treated or control cells in 1.0 ml of nuclear buffer.
2. Centrifuge cells at 2000 *g* for 2 min at 4°C.
3. Resuspend in 250 μl nuclear buffer, transfer to 1.5 ml microfuge tube and centrifuge for 10 s at 4°C in a microfuge at highest speed (12 000 *g*).
4. Resuspend the cells again in 250 μl nuclear buffer and mix with 250 μl of 1.5% LMP agarose with 0.4 mg/ml proteinase K.
5. With a hypodermic needle, inject this mixture of cells and agarose into the glass moulds and leave at 4°C for 30 min for the blocks to solidify.
6. Place the agarose blocks in 3 ml of lysis buffer.
7. Supplement the solution with 50 μl of 2 mg/ml proteinase K and incubate at 37°C for 18 h with gentle mixing on a multipurpose rotator.

Protocol 9. *Continued*

8. After the incubation, rinse the blocks several times in the storage buffer at 4°C for several hours.
9. Agarose blocks can be stored in this buffer for several weeks at 4°C.
10. About 3 mm long plugs of these blocks can be used for electrophoresis.

B. *Field inversion gel electrophoresis*

1. Load the plugs into the slots of the 1.5% agarose gel. If any space is left in the slot between the plug, pour melted LMP agarose and let it solidify at 4°C.
2. Electrophorese at 4°C at 200 V in 0.25 × TAE buffer with the following ramping rate $T_1 = 0.5$ sec, $T_2 = 10$ sec for the first 20 h then $T_1 = 10$ sec, $T_2 = 60$ sec for the next 20 h with a forward to backward ratio of $r = 3$.
3. Stain with ethidium bromide solution (0.5 μg/ml in distilled water).
4. Visualize under UV illumination.
5. Identifiable DNA bands are consistent with cell death by apoptosis.

5.1 Determination of the increased mitochondrial to nuclear DNA ratio for detection and quantitation of apoptosis

Mitochondria and other cellular organelles are preserved in apoptosis, but undergo rapid degradation in necrosis (4). Mitochondria contain multiple copies of a 16.5 kb circular DNA genome which encode several mitochondrial proteins, and mitochondrial ribosomal and transfer RNAs (29, 30). Nuclear DNA (nuDNA) is degraded in all forms of cell death. In contrast, mitochondrial DNA (mtDNA) remains relatively intact in apoptosis, but becomes degraded in necrosis (9, 31). Thus determination of the integrity of any mitochondrial gene relative to the integrity of a nuclear gene provides an accurate procedure for distinguishing apoptosis from necrosis. In addition, the ratio of signal intensity of the mitochondrial to the nuclear gene allows quantitation of the extent of apoptosis in the cell population (9).

DNA extracted as described above (*Protocol 6*) is subjected to restriction enzyme digestion and to Southern blot analysis. The integrity of mitochondrial DNA is determined by the ratio of abundance of a *mt* gene representative of mtDNA to the abundance of a *nu* gene representative of nuDNA. An example of such determination is shown in *Figure 3*, and of the quantitation in *Table 3*.

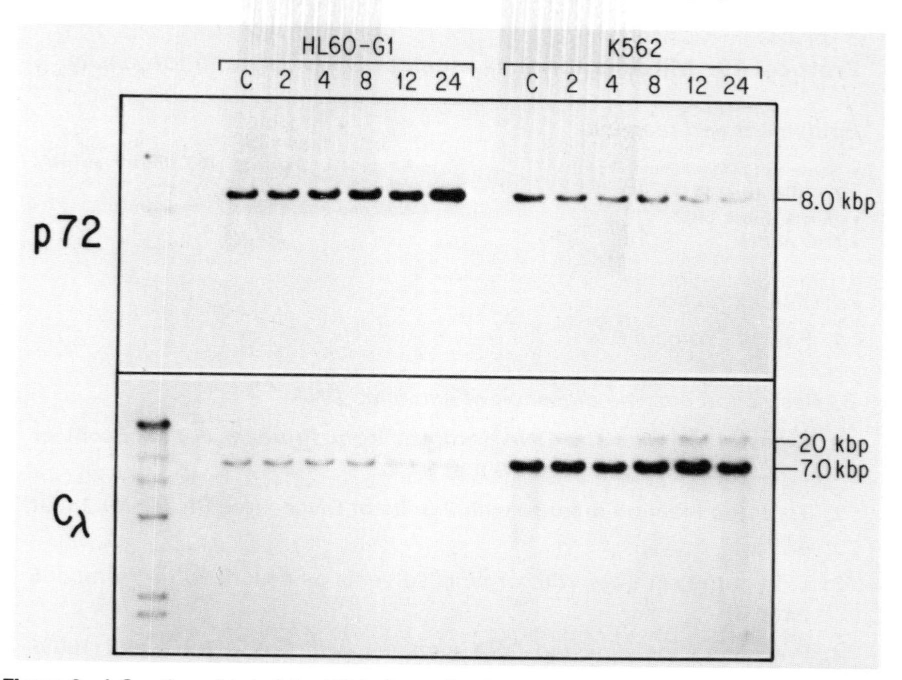

Figure 3. A Southern blot of the DNA shown in *Figure 2* hybridized first to a mitochondrial DNA probe (p72), then re-hybridized to immunoglobulin lambda constant region gene (C_λ). Note that the mitochondrial gene signal increases during doxorubicin-induced apoptosis of HL60 cells, but decreases during doxorubicin-induced necrosis of K562 cells. In contrast, the nuclear gene signal decreases during apoptosis of HL60 cells, but increases slightly during necrosis of K562 cells. The slight increase in nuclear gene signal is due to the enrichment of the DNA sample with nuclear DNA due to the loss of mitochondrial DNA. Reprinted from ref. 9 with permission from Wiley-Liss Inc., a division of John Wiley & Sons, Inc.

Table 3. Ratios of mtDNA to nuDNA during apoptosis in leukaemic cell lines

				p72/c-*myc* ratio[a]	
Cell line	**Treatment**	**Concentration (μM)**	**Duration (h)**	**Eco RI**	**Hind III**
MOLT-4	Control	—	—	1.00	1.00
	Doxorubicin	10	12	3.43 ± 0.56	2.50 ± 0.51
	Dox	10	24	7.50 ± 1.27	7.70 ± 0.92
U937	Control	—	—	1.00	
	Teniposide	5	12	4.18 ± 0.57	ND[b]
	ARA-C	10	12	4.27 ± 0.11	ND

[a] Nuclear DNA and mtDNA levels were compared directly by sequential hybridization of the same membrane with probes for c-myc as a representative nuclear gene, and a probe for mtDNA. Following densitometry of the autoradiographs, the ratios of the relative values of mtDNA/nuclear gene were obtained for untreated cells ('controls') and for each treated group. Control mtDNA/nuDNA ratios were converted to a value of 1.00, and the 'treated' mtDNA/nuDNA ratios were multiplied by the same conversion factor yielding the values presented, ± S.E.M.
[b] Not done.

Protocol 10. Mitochondrial vs. nuclear DNA degradation assay

Equipment and reagents

- Same as *Protocol 6* and *Protocol 7*
- *Eco*RI (Gibco BRL)
- *Hin*dIII (Gibco BRL)
- NaCl, 0.2 M
- Mt-gene probe (e.g., 16S ribosomal RNA, ref. 32)
- Nu-gene probe (e.g., c-*myc*) (Oncor)

A. *DNA extraction*

1. Follow *Protocol 6*.

B. *Restriction enzyme digestion of extracted DNA*

1. Digest 100 μg of the DNA samples from *Protocol 6* with *Eco*RI or *Hin*dIII (3.5 U/μg DNA) for 8 h at 37°C.
2. Treat the reaction mixture with 2 units of DNase-free RNase for 1 h at 37°C.
3. Extract the samples with organic solvents as described in *Protocol 6* (step 3).
4. Precipitate the digested DNA samples with 0.2 M NaCl and 100% ethanol, wash pellets with 70% ethanol, aspirate the supernatant and air dry the pellet.
5. Dissolve the pellet in TE buffer and quantitate using a spectrophotometer.
6. Electrophorese 10 μg of the DNA sample as in *Protocol 8*.
7. Depurinate and denature the samples in the gel as described in ref. 33.
8. Southern transfer the DNA as described by Southern (34) and immobilize the DNA on the membrane by drying and baking at 80°C for 2–5 h.
9. Nick translate the probes as described in ref. 35.
10. Add probes to the hybridization buffer.
11. Wash the membranes in SSC solution.
12. Wrap the membranes in plastic and expose to autoradiographic film at −80°C for variable periods of time.
13. Analyse the film by densitometric image analysis for intensity.

6. Assessment of nuclear proteolytic activity in cells undergoing apoptosis

Recent studies have suggested that proteases may be involved in the initiation of apoptosis (10, 12). The following protocol has proven effective in analysing the fate of selected polypeptides during chemotherapy-induced apoptosis in human tissue culture cell lines (10, 36), and can be used to monitor this process. For instance, ICE may be the protease which, directly or indirectly, activates the apoptotic endonuclease (12), and proteolytic cleavage of poly(ADP-ribose) polymerase has been shown to be an early marker of apoptosis (10). The reagents for these reactions are not currently easily available, but apoptosis can be monitored by proteolytic digestion of nuclear proteins such as lamin B or topoisomerase II, since proteolysis of these proteins also occurs early in apoptosis (36).

Protocol 11. Demonstration of nuclear protein fragmentation

Equipment and reagents

- Solubilization buffer: 6 M guanidine hydrocholoride, 250 mM Tris–HCl (pH 8.5), 10 mM EDTA. Add protease inhibitor α-phenylmethylsulphonyl fluoride (PMSF, final concentration 1 mM) and reducing agent β-mercaptoethanol (final concentration 1% v/v) to each aliquot of solubilization buffer immediately before use
- 8 M urea (deionized over mixed-bed resin before use)
- Reagents for SDS–polyacrylamide gel electrophoresis and Western blotting (37, 38)
- Antibodies against the desired protein. Antibodies against topoisomerase II are available from Topogen. Polyclonal and monoclonal antibodies against the nuclear envelope lamins are also widely available

Method

1. Incubate cells with the apoptosis-inducing agent for the desired length of time.
2. Remove tissue culture medium. Wash cells once with serum-free buffer (e.g., serum-free RPMI 1640 medium or phosphate-buffered saline).
3. Solubilize cells in solubilization buffer containing PMSF and β-mercaptoethanol. A total volume of 1 ml for cell pellets or 3 ml for a 100 nm diameter dish of cells is convenient.
4. Sonicate the sample to shear the DNA.
5. After a ⩾ 4 h incubation at room temperature to ensure complete reduction of the sample, react with iodoacetamide (150 mM final concentration) for 1 h at room temperature to block sulphydryl groups. Stop the reaction by adding additional β-mercaptoethanol (10 μl/ml solution).
6. Dialyse the sample at 40°C against 4 M urea (multiple passes) and then against 0.1% (w/v) SDS.

Protocol 11. *Continued*

7. After removing an aliquot of the dialysed sample for protein determination (39), lyophilize sample to dryness and reconstitute in the desired volume of electrophoresis sample buffer. Samples at 2–5 $\times$ 10^4 cells/μl or 2–5 μg/μl are convenient.
8. Perform SDS–polyacrylamide gel electrophoresis, electrophoretic transfer and Western blotting using standard techniques (37, 38).
9. Compare band intensity with signal for a nuclear protein resistant to degradation, e.g. histone H1.

7. Concluding remarks

This chapter should be consulted together with Chapter 8, which describes another major group of procedures for the detection and quantitation of apoptosis, namely the applications of flow cytometry for this purpose, and with Sections 5.7 and 6.5 in Chapter 6, which deal with filter binding assays for detection and quantitation of apoptosis. Additional background information on the various forms of cell death is also summarized there. While comparing these chapters it should be remembered that the concept of apoptosis is still an amorphous entity, and it is therefore legitimate at this time to disagree on what constitutes apoptosis as opposed to the other forms of cell death.

Acknowledgements

We thank our colleagues for helpful comments on the manuscript. Experimental work presented here was supported by USPHS grant CA-44722 from the National Cancer Institute.

References

1. Sulston, J. E., Schierenberg, E., White, J. G., and Thomson, J. N. (1983). *Dev. Biol.,* **100**, 64.
2. Wyllie, A. H., Morris, R. G., Smith, A. L., and Dunlop, D. (1984). *J. Pathol.,* **142**, 67.
3. Eastman, A., Grant, S., Lock, R., Tritton, T., VanHouten, N., and Yuan, J. (1994). *Cancer Res.,* **54**, 2812.
4. Kerr, J. F. R., Wylie, A. H., and Currie, A. R. (1972). *Br. J. Cancer,* **26**, 239.
5. Skalka, M., Matyasova, J., and Cejkova, M. (1976). *FEBS Lett.,* **72**, 271.
6. Wylie, A. H. (1980). *Nature*, **284**, 555.
7. Oberhamner, F., Wilson, J. W., Dive, C., Morris, I. D., Hickman, J. A., Wakeling, A. E., Walker, P. R., and Sikorska, M. (1993). *EMBO J.,* **12**, 3679.

8. Walker, R. P., Weaver, V. M., Lach, B., Leblanc, J., and Sikorska, M. (1994). *Exp. Cell Res.,* **213**, 100.
9. Tepper, C. G. and Studzinski, G. P. (1993). *J. Cell. Biochem.,* **52**, 352.
10. Kaufmann, S. H., Desnoyers, S., Ottaviano, Y., Davidson, N. E., and Poirier, G. G. (1993). *Cancer Res.,* **53**, 3976.
11. Ellis, H. M. and Horvitz, R. (1986). *Cell,* **44**, 817.
12. Yuan, J., Shaham, S., Ledoux, S., Ellis, H. M., and Horvitz, H. R. (1993). *Cell,* **75**, 641.
13. Wong, P., Pineault, J., Lakins, J., Taillefer, D., Leger, J., Wang, C., and Tenniswood, M. (1993). *J. Biol. Chem.*, **268**, 5021.
14. Itoh, N., Yonehara, S., Ishii, A., Yonehara, M., Mizushima, S., Ssmeshima, M., Hase, A., Seto, Y., and Nagata, S. (1991). *Cell,* **6**, 233.
15. Oehm, A., Behrmann, I., Falk, W., Pawlita, M., Maier, G., Klas, C., Li-Weber, M., Richards, S., Dhein, J., Tranth, B. C., *et al.* (1992). *J. Biol. Chem.*, **267**, 10709.
16. Wyllie, A. H., Kerr, J. F. R., and Currie, A. R. (1980). *Int. Rev. Cytol.*, **68**, 251.
17. Arends, M. J., Morris, R. G., and Wyllie, A. H. (1990). *Am. J. Pathol.*, **136**, 593.
18. Xu, H.-M., Tepper, G. C., Jones, J. B., Fernandez, C. E., and Studzinski, G. P. (1993). *Exp. Cell Res.*, **209**, 367.
19. Hayashi, M., Morita, T., Kodama, Y., Sofuni, T., and Ishidate, M. (1990). *Mutat. Res.*, **245**, 245.
20. Carson, F. L. (1990). *Histotechnology: A Self Instructional Text*, ASCP Press, Chicago.
21. Gavriel, Y., Sherman, Y., and Ben-Sasson, S. A. (1992). *J. Cell Biol.*, **119**, 493.
22. Wijsman, J. H., Jonker, R. R., Keijzer, R., VanDeVelde, C. J. H., Cornelisse, C. J., and Van Dierendonck, J. H. (1993). *J. Histochem. Cytochem.*, **41**, 7.
23. Catchpoole, D. R. and Stewart, B. W. (1993) *Cancer Res.*, **53**, 4287.
24. Falcieri, E., Martelli, A. M., Bareggi, R., Cataldi, A., and Cocco, L. (1993). *Biochem. Biophys. Res. Commun.*, **193**, 19.
25. Wagner, A. J., Small, M. B., and Hay, N. (1993). *Mol. Cell. Biol.*, **13**, 2432.
26. Gong, J., Traganes, F., and Darzynkiewicz, Z. (1994). *Anal. Biochem.*, **218**, 314.
27. Brown, D. G., Sun, X. M., and Cohen, G. M. (1992). *J. Biol. Chem.,* **268**, 3037.
28. Filipski, J., Leblanc, J., Youdale, T., Sikorska, M., and Walker, P. R. (1990). *EMBO J.,* **9**, 1319.
29. Anderson, S., Bankier, A. T., Barrell, B. G., deBruijn, M. H. L., Coulson, A. R., Drouin, J., Eperon, I. C., Nierlich, D. P., Roe, B. A., Sanger, F., Schreier, P. H., Smith, A. J. H., Staden, R., and Young. I. G. (1981). *Nature,* **290**, 457.
30. Clayton, D. A. (1984). *Annu. Rev. Biochem.,* **53**, 573.
31. Murgia, M., Pizzo, P., Sandonia, D., Zanovello, P., Rizzuto, R., and DiVirgilio, F. (1992). *J. Biol. Chem.,* **267**, 10939.
32. Tepper, C. G., Pater, M. M., Pater, A., Xu, H. M., and Studzinski, G. P. (1992). *Anal. Biochem.,* **203**, 127.
33. Wahl, G. M., Stern, M., and Stark, G. R. (1979). *Proc. Natl. Acad. Sci. USA,* **76**, 3683.
34. Southern, E. M. (1975). *J. Mol. Biol.,* **98**, 503.
35. Maniatis, T., Fritsch, E. F., and Sambrook, J. (1982). *Molecular Cloning: A Laboratory Manual*. Cold Spring Harbor Lab., New York.

36. Kaufmann, S. H. (1989). *Cancer Res.,* **49**, 5870.
37. Laemmli, U. K. (1985). *Nature,* **227**, 680.
38. Kaufmann, S. H. and Shaper, J. H. (1992). *Methods Mol. Biol.,* **10**, 235.
39. Smith, P. K., Krohn, R. I., Hermanson, G. T., Mallia, A. K., Gartner, F. H., Provenzano, M. D., Fujimoto, E. K., Goeke, N. M., Olson, B. J., and Klenk, D. C. (1985). *Anal. Biochem.,* **150**, 76.

8

Analysis of cell death by flow cytometry

ZBIGNIEW DARZYNKIEWICZ, XUN LI, JIANPING GONG, SHINSUKE HARA, and FRANK TRAGANOS

1. Introduction

Flow cytometry allows one rapidly and accurately to measure individual cells in large cell populations. Because several cell attributes can be measured simultaneously and the data are recorded in a mode that allows one to relate all the measured attributes to a particular cell ('list mode'), their multivariate analysis provides information on the relationship between these attributes. In contrast to biochemical analysis in bulk, cellular heterogeneity can be estimated and cell populations with distinct characteristics can be discriminated. If needed, cells with particular features, can be electronically sorted. These advantages have all contributed to the fact that flow cytometry has become a methodology of choice in a variety of functional and structural assays of the cell, including cell viability.

Several flow cytometric assays of cell viability are presented in this chapter. Nearly all these methods have been used, or are in routine use, in our laboratory, and several were developed or modified by us. Because at least two distinct modes (mechanisms) of cell death can be distinguished, and there are differences between the cells at early and late stages of death, each of the assays is evaluated with respect to its usefulness in discriminating the mode and the stage of cell death. Comparison of the methods in terms of their simplicity, specificity in discriminating between modes of cell death, data interpretation, cost, and usefulness in particular types of applications are also discussed. Because of the space limitation in this volume, some methods had to be omitted; they are presented in full elsewhere (1).

2. Modes of cell death

Based on distinct differences in morphological, biochemical and molecular changes of the dying cell, two modes of cell death have been described: apoptosis and necrosis (reviews, 2–5). The third mode, so-called mitotic death or delayed reproductive death is less well characterized (6, 7); however, because it shows some similarity to apoptosis, it is frequently depicted as the latter. Several assays of cell viability described in this chapter can be used to discriminate between these modes of cell death. It should be stressed, however, that on some occasions, depending on the cell type and nature of the factor affecting cell viability, the mode of cell death may not be typical of either necrosis or apoptosis, lacking the characteristic features of either of these mechanisms (e.g. 8, 9).

2.1 Apoptosis

Apoptosis is a particular mode of cell death which can be distinguished by a characteristic pattern of morphological and molecular changes (Chapter 7, also reviews, refs 2–5). Due to rapid cell dehydration, cells which were originally round often become elongated or convoluted in shape and diminished in size. Chromatin condensation and the loss of distinct chromatin structure, which occurs in parallel with cell shrinkage, starts at the nuclear periphery and is followed by nuclear fragmentation. Distinct hyperchromicity and homogeneity characterize the stainability of DNA with such fluorochromes as 4,6-diamidino-2-phenylindole (DAPI), Hoechst 33342, propidium iodide or (PI), in the fragmented nuclei. Nuclear fragments, together with the constituents of the cytoplasm (including intact organelles), are then packaged into so-called apoptotic bodies, which, enveloped in plasma membrane, detach from the dying cell and, *in vivo*, are phagocytosed by neighbouring cells without invoking an inflammatory response.

Increased cytoplasmic Ca^{2+} concentration and activation of proteases, as well as one, or several, endonuclease(s) which degrade DNA at the internucleosomal (linker) sections are also very characteristic events of apoptosis (2–5). Thus, the products of DNA degradation are discontinuous and of the size of nucleosomal and oligonucleosomal DNA sections that generate a very characteristic 'ladder' pattern during gel electrophoresis. Also typical is activation of some genes; the latter suggests that products of such genes are required for apoptosis to occur. However, exceptions are noted, and several cell types (so called apoptosis-primed cells) may undergo apoptosis when protein synthesis is inhibited, e.g. in the presence of cycloheximide. Even during advanced stages of apoptosis, the structural and functional integrity of the plasma membrane, mitochondria, and lysosomes, are generally preserved (reviews, 2–5). Thus, considering all the changes that occur in a cell undergoing apoptosis, the most characteristic feature of this process is active participation of the cell in its demise. The cell activates a cascade of molecular events that result in orderly degradation of the cell constituents with minimal impact on the neighbouring tissue.

The long-standing interest in apoptosis in disciplines such as embryology, immunology, or endocrinology stems from the observations that this mode of cell death plays a pivotal role in tissue development, organ involution and immunity (reviews, in ref. 5). Apoptosis affecting $CD4^+$ lymphocytes of HIV-infected patients also appears to play a major role in the pathogenesis of AIDS (10).

The interest in apoptosis in oncology is more recent. It derives from several findings in different fields. Thus, it has been repeatedly observed that this mode of cell death in tumours is triggered by pharmacological doses of anti-cancer drugs of various classes, radiation or hyperthermia (11). Furthermore, it has been noted that the cell's ability to undergo apoptosis plays a role in development of B-cell lymphomas (12), whereas the frequency of spontaneous apoptosis is associated with progression of tumour malignancy (13). It has also been observed that the inherent capacity of tumour cells to respond by apoptosis correlates with expression of several oncogenes or tumour suppressor genes such as *bcl*-2, c-*myc*, *ras*, or *p53*, and may be prognostic of the treatment (2–5, 14). Research is underway in many laboratories to find the means to modify the responsiveness of normal and tumour cells to various agents by apoptosis. It is expected that a knowledge of the molecular mechanisms of apoptosis will be helpful in developing new anti-tumour strategies.

2.2 Necrosis

Whereas an active participation of the affected cell, often involving *de novo* protein synthesis, is an essential feature of apoptosis, necrosis is a passive and degenerative process. The early event of necrosis is swelling of cell mitochondria, followed by rupture of the plasma membrane and release of the cytoplasmic contents (reviews in ref. 5). Because of the release of cell constituents, which include many proteolytic enzymes, necrosis, in contrast to apoptosis, triggers an inflammatory reaction in the tissue, often resulting in scar formation. DNA degradation is not so extensive as in the case of apoptosis, and the products of degradation are heterogeneous in size, and do not form any discrete bands during gel electrophoresis.

Necrosis generally represents a cell's response to gross injury and is frequently induced by an overdose of cytotoxic agents. It has been observed, however, that certain cell types do respond even to pharmacological concentrations of some drugs or moderate doses of physical agents by necrosis rather than apoptosis though the reason for the difference in response is not entirely clear.

2.3 Atypical apoptosis: delayed reproductive, or mitotic, cell death

Numerous examples of cell death have been described in which the pattern of morphological and/or biochemical changes resembled neither typical apoptosis nor necrosis but exhibited some features of both (e.g. refs 6–9). In some cases, the integrity of the plasma membrane was preserved but DNA degradation was random, without evidence of any nucleosomal pattern. In other situations, DNA degradation was typical of apoptosis but nuclear fragmentation was not apparent. It is likely that, in such cases, one or more of the many branches (pathways) of the apoptotic cascade (e.g. activation of apoptosis-associated endonuclease) were inhibited, resulting in cell death that showed only a few features of apoptosis and, thus, could not be identified as such.

So-called delayed reproductive, or mitotic cell, death are the terms used to describe death that occurs as a result of exposure to relatively low doses of drugs or radiation, which induce irreparable damage, but allow cells to complete at least one round of cell division (6, 7). Because the dying cells show some features of apoptosis, it is possible that 'delayed reproductive death' is actually delayed apoptosis, whose picture is complicated by the secondary changes in cell metabolism, such as related to growth imbalance, perturbed cell cycle progression, etc.

3. Features discriminating dead cells

3.1 Structural and functional integrity of the plasma membrane

One of the major features distinguishing dead from live cells is the loss of transport function and often even the loss of the structural integrity of the plasma membrane. Numerous assays of cell viability have been developed based on this phenomenon. Because the intact membrane of live cells excludes charged dyes such as Trypan Blue or PI, short-term incubation with these dyes results in selective labelling of dead cells, whereas live cells show no, or minimal dye uptake (15). The PI exclusion assay is most commonly used in flow cytometry. A short (5–10 min) incubation in the presence of 10–20 μg/ml of PI, in isotonic media, labels dead cells, that cannot exclude the dye. After crossing the plasma membrane PI binds to DNA and dsRNA by intercalation, which leads to intense fluorescence of this dye, which otherwise, unbound and in aqueous solution, shows weak fluorescence.

Another assay of membrane integrity employs the non-fluorescent esterase substrate, fluorescein diacetate (FDA). This substrate, after being taken up by live cells is hydrolysed. The product of the hydrolysis, fluorescein, is highly fluorescent, emitting green fluorescence (16). Incubation of cells in the presence of both PI and FDA labels live cells green (fluorescein) and dead cells red (PI). This is a convenient assay, widely used in flow cytometry.

The integrity of the plasma membrane can also be probed by assaying cell resistance to enzymes such as trypsin and DNase. Whereas live cells remain intact during incubation with these enzymes, the cells with a damaged plasma membrane are nearly totally digested and removed from the analysis (17, 18).

3.2. Cell organelles

Several assays of cell viability are based on the functional tests of cell organelles. For example, the cationic dye rhodamine 123 (Rh123) which fluoresces green, accumulates in mitochondria of live cells due to the mitochondrial transmembrane potential (19). Cell incubation with Rh123 results in labelling of live cells whereas dead cells show minimal Rh123 uptake. Cell incubation with both Rh123 and PI labels live cells green (Rh123) and dead cells red (20). A transient phase of cell death, however, is observed, when the cells partially lose ability to exclude PI and yet stain intensively, even more than live, intact cells, with Rh123 (20). This may suggest that the mitochondrial transmembrane potential is temporarily increased, concomitant with a loss of plasma membrane function.

Another organelle which can be probed is the lysosome. Incubation of cells in the presence of 1–2 μg/ml of the metachromatic dye Acridine Orange (AO) results in uptake of this dye by lysosomes of live cells which fluoresce red (21). The uptake is the result of an active proton pump in lysosomes: the high proton concentration (low pH) causes AO, which can enter the lysosome in the uncharged form, to become protonated and thus entrapped in the organelle. Dead cells, at that low AO concentration, exhibit weak green and minimal red fluorescence. This assay is especially useful for cells that have numerous active lysosomes, such as monocytes, macrophages, etc.

In cells undergoing apoptosis, especially at early phases of this process, the plasma membrane is preserved and most organelles and cellular functions remain relatively unchanged compared to live cells. Therefore, a simple discrimination of live and dead cells by dye exclusion, or based on functional assays of some organelles, is not always adequate to identify cells that die by apoptosis. Other assays, therefore, have to be used in such situations.

3.3 Specific features of apoptotic cells

Extensive DNA cleavage and preservation of at least some functions of the cell membrane provide the basis for development of most flow cytometric assays to distinguish apoptotic cells. Two different approaches are most frequently used to detect DNA cleavage. The first is based on the extraction of the degraded, low molecular-weight DNA from ethanol-prefixed, or detergent-permeabilized cells, and subsequent cell staining with DNA fluorochromes. Apoptotic cells then show reduced DNA stainability (content of high molecular-weight DNA) (review, ref. 21). In the second approach, the numerous DNA strand breaks in apoptotic cells can be labelled with biotinylated or digoxygenin-conjugated nucleosides in a reaction employing exogenous terminal deoxynucleotidyl transferase (TdT) or DNA polymerase (nick translation) (22, 23). These procedures are described in Chapter 7.

Chromatin condensation is also a quite specific feature of apoptotic cells.

Chromatin condensation can be assayed by DNA *in situ* sensitivity to denaturation. DNA in condensed chromatin is markedly more sensitive to denaturation than is DNA in non-condensed chromatin (24).

Transport of the benzimidazole dye Hoechst 33342 (HO342) across the plasma membrane is altered in apoptotic cells. Differential staining of DNA with this dye, thus, discriminates between live and apoptotic cells (25, 26).

Practical aspects of several of the cell viability assays, their specificity in discriminating between apoptosis and necrosis and applicability, will be described below.

4. Light scattering properties of dying cells

The cell's capability to scatter light changes during death reflecting the morphological changes, namely chromatin condensation, nuclear fragmentation, cell shrinkage, and shedding of apoptotic bodies. Thus, the measurement of light scatter is one of the simplest assays of cell viability. Reduced ability to scatter light in the forward direction and either an increase (27) or no change in the 90° light scatter (21) characterize cells in the early phase of apoptosis. Later, both the forward and right angle light scatter signals are decreased.

Assay of cell viability by light scatter measurement is very simple and can be combined with analysis of surface immunofluorescence, for example to identify the phenotype of the dying cell. It can also be combined with functional assays such as mitochondrial potential, lysosomal proton pump, exclusion of PI or plasma membrane permeability to such dyes as Hoechst 33342.

The light scatter changes, however, are not specific to apoptosis because mechanically broken cells, isolated cell nuclei and necrotic cells also have low light scatter properties. Furthermore, a loss of cell surface antigen may accompany apoptosis, especially at later stages of cell death and this may complicate the bivariate light scatter/surface immunofluorescence analysis. The light scatter measurement, therefore, requires several controls, and should be accompanied by another, more specific assay, or at least by confirmation of apoptosis by microscopy.

5. Removal of dead cells by incubation with DNase I and trypsin

Limited exposure of live cells sequentially to DNase I and trypsin has little effect on their morphology, function or viability. On the other hand, cells with a damaged plasma membrane are digested by these enzymes to the degree that they can be gated out during the measurement, e.g. by raising the triggering threshold of the light scatter, DNA or protein fluorescence signals (17, 18).

The procedure described in *Protocol 1* removes cells with impaired integrity of the plasma membrane from suspension; all remaining cells generally

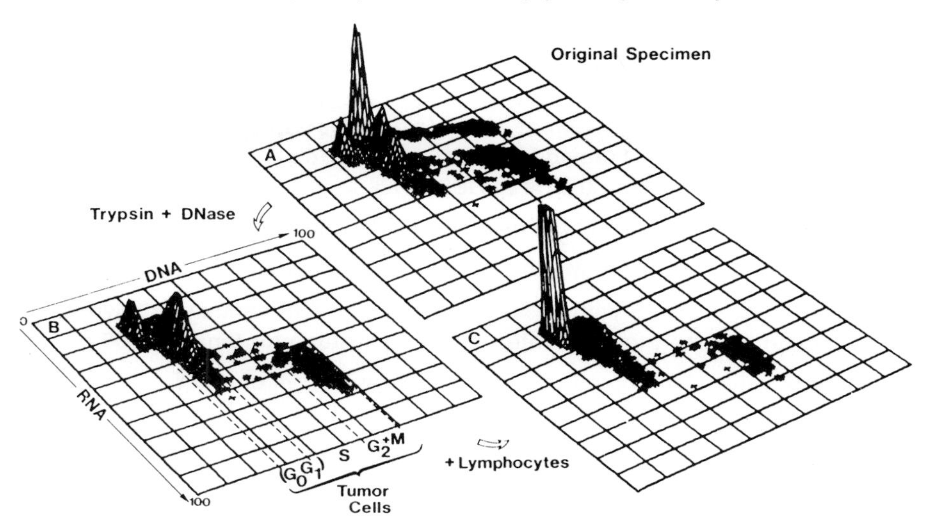

Figure 1. The DNA and RNA (Acridine Orange) distribution of bone marrow from a patient with resistant neuroblastoma. The original sample (A) contained two populations with increased DNA content but different RNA (low and high) content. Following incubation with trypsin and DNase, the aneuploid population with low RNA content disappeared (B). When normal peripheral blood lymphocytes were admixed with the sample (C), the population with low DNA (diploid cells) was enriched indicating that the tumour was indeed hyperdiploid with a DNA index of 2.65.

exclude trypan blue or PI. However, because trypsin digests the cell coat, epitopes of many cell surface antigens of live cells may not be preserved. Incubation of enzyme-treated cells in culture medium at 37°C for several hours may restore most antigens.

Application of *Protocol 1* selectively removes isolated nuclei, necrotic and very late apoptotic cells, as well as cells that are mechanically damaged (e.g. by sonication, syringing, etc.) but not live and early apoptotic cells. *Figure 1* illustrates an example in which an aneuploid tumour (neuroblastoma) contained two cell populations with differing RNA content. Generally, very low RNA content by this staining technique, may indicate loss of cytoplasmic material. Trypsin + DNase treatment completely abolished the low RNA content population and improved identification of the diploid and aneuploid populations.

Protocol 1. Trypsin–DNase I digestion of damaged cells

Working enzyme solutions

- DNase: dissolve 1 mg of DNase I in 1 ml of Hank's buffered salt solution (HBSS; with Mg^{2+} and Ca^{2+}). Aliquots of 100 μl of DNase I solution in microfuge tubes may be stored at −40°C
- Trypsin: dissolve 20 mg of trypsin in 1 ml of HBSS. Aliquots of 100 μl of trypsin may be stored at −40°C in microfuge tubes

Protocol 1. *Continued*

Method

1. Centrifuge cells for 5 min at 100 *g* (e.g. dispersed mechanically from the solid tumour, obtained from tissue culture, etc.). Suspend cell pellet (10^6–5×10^6 cells) in 0.8 ml of HBSS.
2. Add 100 μl of DNase I, and incubate for 15 min at 37°C.
3. Add 100 μl of trypsin and incubate for an additional 30 min.
4. Add 5 ml of HBSS with 10% calf serum or HBSS containing soybean trypsin inhibitor, to inactivate trypsin. Centrifuge for 5 min at 100 *g*.
5. Rinse cells once in 5 ml of HBSS, suspend in 1 ml of HBSS.
6. Fix cells in suspension (e.g. in 10 ml of 70% ethanol), or stain with a desired fluorochrome (e.g. after permeabilization with 0.1% Triton X100 in the presence of 1% albumin) shortly after this procedure.

6. Exclusion of PI combined with hydrolysis of fluorescein diacetate (FDA)

FDA, a substrate for esterases, can penetrate into live cells. The product of the hydrolysis, fluorescein, is highly fluorescent and charged. Because of the charge, it becomes entrapped in the cell. This assay (*Protocol 2*), which combines counterstaining of dead cells with PI, has been widely used in classical cytochemistry and in flow cytometry (16).

The method described in *Protocol 2*, which is based on analysis of the integrity of the plasma membrane, as other assays based on the same principle (e.g. *Protocol 1*), discriminates between necrotic cells (or cells with a damaged membrane) and live cells but not between live and early apoptotic cells.

Protocol 2. Exclusion of PI coupled with hydrolysis of FDA

Staining solutions

- FDA stock solution: dissolve 1 mg of FDA in 1 ml of acetone
- PI stock solution: dissolve 1 mg of PI in 1 ml of distilled water

Method

1. Suspend approximately 10^6 cells in 1 ml of HBSS.
2. Add 2 μl of the FDA stock solution.
3. Incubate cells at 37°C for 15 min.

4. Add 20 μl of the PI stock solution and incubate for 5 min at room temperature.
5. Analyse by flow cytometry. Live cells exhibit green, dead cells, red fluorescence.
   - Use excitation in blue light (e.g. 488 nm line of the argon ion laser).
   - Measure green fluorescence at 530 ± 20 nm.
   - Measure red fluorescence at > 620 nm.

7. Uptake of Rh123 combined with exclusion of PI

As mentioned, when live cells are exposed to the cationic fluorochrome Rh123, the dye is rapidly accumulated in mitochondria due to the transmembrane potential of these organelles (19). Live cells with an intact plasma membrane and active (charged) mitochondria concentrate this dye and exhibit strong green fluorescence. In contrast, dead cells stain weakly with this dye. Thus, dual staining with Rh123 and PI was proposed as a cell viability test (20). This is also an assay of the mitochondrial transmembrane potential.

When stained as described in *Protocol 3*, live and early apoptotic cells stain green with Rh123 and exclude PI. Dead cells, which have damaged mitochondrial or plasma membrane (e.g. necrotic cells), have minimal green fluorescence but stain with PI (*Figure 2*). During the early phase of cell death the cells may stain intensely with both dyes (20).

Protocol 3. Exclusion of PI coupled with Rh123 staining

Staining solutions

- Rh123: dissolve 1 mg of Rh123 (Molecular Probes) in 1 ml of distilled water (stock solution)
- PI: dissolve 1 mg of PI in 1 ml of distilled water (stock solution)

Method

1. Add 5 μl of Rh123 stock solution to approximately 10^6 cells suspended in 1 ml of tissue culture medium (or HBSS) and incubate for 5 min at 37°C.
2. Add 20 μl of the PI stock solution and keep for 5 min at room temperature.
3. Analyse by flow cytometry.
   - Use excitation with blue light (e.g. 488 nm laser line).
   - Measure the Rh123 fluorescence in green wavelength (530 ± 20 nm).
   - Measure PI red fluorescence, as described in *Protocol 2*.

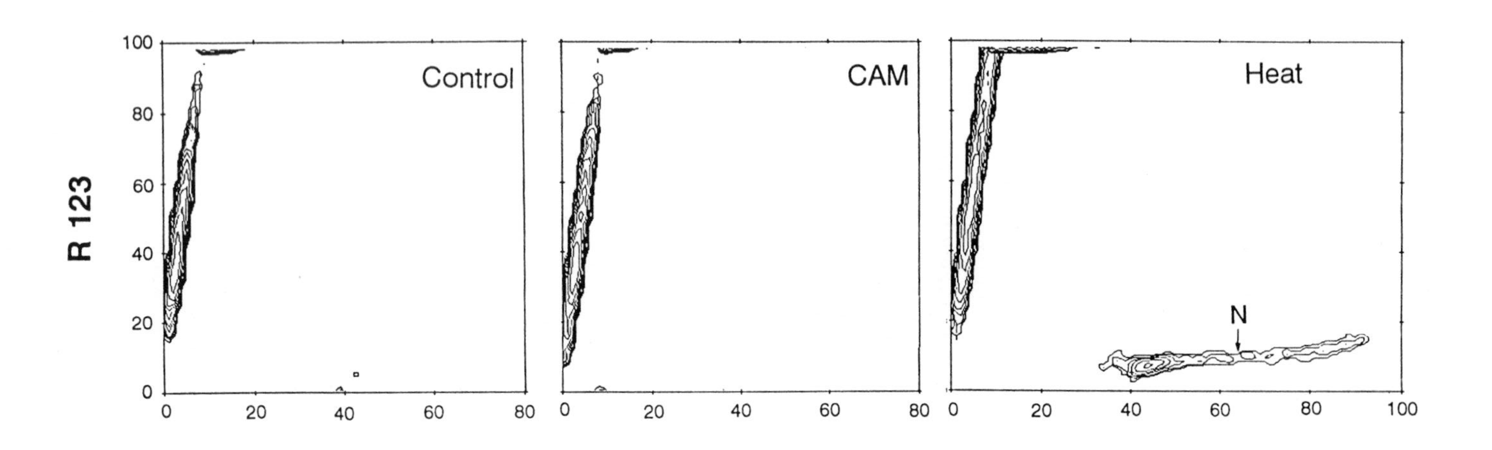
Control
CAM
Heat
N
R 123
Propidium Iodide
0
20
40
60
80
100

Figure 2. Stainability of live, apoptotic and necrotic cells with Rh123 and PI. The mitochondrial probe Rh123 (green fluorescence) is taken up by live cells from control HL-60 cultures (Control); live cells exclude PI. Induction of apoptosis in approximately 40% of the cells by incubation with 0.15 μM camptothecin for 4 h (CAM) neither changes Rh123 uptake nor affects exclusion of PI. Cell heating (45°C, 2 h) causes necrosis (Heat). Early necrotic cells have elevated Rh123 uptake but exclude PI (not shown). Late necrotic cells (N) have minimial Rh123 fluorescence and stain intensely with PI. (Reprinted with permission from ref. 21.)

8. Exclusion of PI followed by counterstaining with HO342

The charged dye PI, as mentioned, is excluded by live cells, whereas HO342 penetrates through the plasma membrane and can stain DNA in live cells. The method combining these dyes was proposed by Stohr and Vogt-Schaden (28) and modified by several authors (29, 30). In this assay, the cells are initially exposed to PI and then stained with HO342. Compared with live cells, HO342 fluorescence is suppressed in dead cells. The latter, however, stain more intensely with PI. This method, a modification of that of Pollack and Ciancio (30), is described in *Protocol 4*.

Protocol 4. PI exclusion coupled with HO342 staining

Stock solutions and reagents

- HO342 (stock solution): dissolve 3 mg of HO342 (Molecular Probes) in 10 ml of distilled water
- HO342 (working solution): dilute the HO342 stock solution 1:4 in PBS (Ca^{2+} and Mg^{2+} free)
- PI (working solution): dissolve 2 mg of PI in 100 ml of PBS
- Fixative: 25% (v/v) ethanol in PBS

Method

1. Centrifuge 3×10^5–10^6 cells, decant the medium and vortex the pellet.
2. Add 100 μl of PI working solution, vortex and keep on ice for 30 min.
3. Add 1.9 ml of fixative (25% ethanol) and vortex.
4. Add 50 μl of HO342 working solution, vortex and keep on ice for at least 30 min. Samples are stable for up to 3 days in this solution, when kept at 0–4°C.
5. Analyse by flow cytometry.
   - Excitation of HO342 is with UV light, with a maximum at 340 nm.
   - PI also has an absorption band in the UV light spectrum.

Protocol 4. *Continued*

- Excitation of both PI and HO342 can, thus, be achieved using the 351 nm line of the argon ion laser, or in the case of illumination with high pressure mercury lamp, using UG1 filter.
- Measure the blue fluorescence of HO342; use a combination of filters and dichroic mirrors to obtain maximum transmission at 460±20 nm.
- Measure the red fluorescence of PI with a long pass filter at > 620 nm.

Figure 3 illustrates the positions of live, apoptotic and necrotic cells following cell staining with PI and HO342. As is evident, the early apoptotic cells (Ap 1) have lowered stainability with HO342 compared to live cells from the control culture. The loss of HO342 stainability is likely to be the result of DNA degradation and extraction of low molecular-weight DNA from the cells following their permeabilization with ethanol. PI fluorescence of Ap cells is also low. Loss of the plasma membrane function late in apoptosis results in increased stainability with PI (Ap 2). Necrotic cells, or cells with

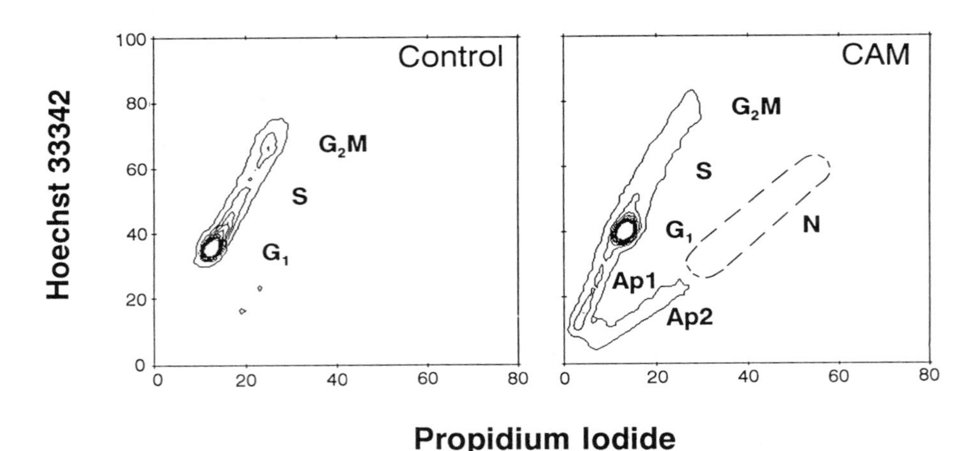

Figure 3. Stainability of apoptotic and necrotic cells with PI and HO342 under conditions when HO342 is applied after cell exposure to PI and following their permeabilization. HL-60 cells, untreated (Control) and treated with 0.15 μM camptothecin (CAM) for 6 h to induce apoptosis (44), were incubated with 10 μg/ml of PI, then permeabilized with 25% ethanol and stained with HO342, as originally described by Pollack and Ciancio (30). In this assay, the cells with an undamaged plasma membrane exclude PI and stain predominantly blue with HO342. The cells that cannot exclude PI have more intense PI fluorescence and proportionally lower HO342 fluorescence. The live cells from the control culture stain strongly with HO342, in proportion to their DNA content. The early apoptotic cells from the CAM-treated cultures show diminished HO342 fluorescence (Ap1). Late apoptotic cells (Ap2) cannot exclude PI and, thus, stain more intensely red. The position of necrotic cells (N), which also cannot exclude PI but have higher DNA content than Ap2, is indicated by the broken outline. (Reprinted with permission from ref. 21.)

damaged plasma membrane function (e.g. by repeated freezing and thawing) have high PI and low HO342 fluorescence.

9. Uptake of HO342 combined with exclusion of PI

Ormerod *et al.* (26) and Dive *et al.* (25) have reported that a short exposure of apoptotic cells to HO342 labels them strongly with this fluorochrome, perhaps as a result of increased permeability to this dye. This is in contrast to live, non-apoptotic cells which require much longer incubation with HO342 to obtain a comparable intensity of fluorescence. Supravital uptake of HO342 combined with exclusion of PI and analysis of the cell's light scatter properties, proposed by these authors, provides an attractive assay of apoptosis (25, 26).

It should be emphasized that although the same dyes (HO342 and PI) are used, the principle of this method is entirely different from the method presented in *Protocol 4*. This method (*Protocol 5*) identifies apoptotic cells exclusively based on their altered plasma membrane permeability to HO342. At the time of the preparation of this chapter, it was not known whether this change in membrane permeability was generic, e.g. whether it applied to different cell types or different agents that trigger apoptosis.

When stained as in *Protocol 5*, apoptotic cells have increased blue HO342 fluorescence compared to live, non-apoptotic cells. However, the intensity of the cell's blue fluorescence changes with the time of incubation with HO342. An optimal incubation time (generally 2–10 min) should be found for best discrimination of apoptotic cells. The decrease in light scatter, combined with HO342 uptake, is helpful in identifying apoptotic cells. Necrotic cells are counterstained with PI and fluoresce red.

Protocol 5. Uptake of HO342 coupled with PI exclusion

Reagents

- HO342 stock solution: 0.1 mg of HO342 in 1 ml of distilled water
- PI stock solution: 1 mg of PI in 1 ml of distilled water

Method

1. Suspend approximately 10^6 cells in 1 ml of culture medium, with 10% serum.
2. Add 10 μl of stock solution of HO342.
3. Incubate cells for 2–7 min at 37°C.
4. Cool the sample on ice, centrifuge for 5 min at 100 *g*.
5. Resuspend the cell pellet in 1 ml of PBS.
6. Add 5 μl of the PI stock solution.

Protocol 5. *Continued*

7. Analyse by flow cytometry; the analysis conditions (the excitation and emission spectra of the dyes) are as in *Protocol 4*.

10. Extraction of the degraded DNA from apoptotic cells

As mentioned, one of the early and relatively specific events of apoptosis is activation of an endonuclease which nicks DNA preferentially at the internucleosomal (linker) sections (2–4). Fixation of cells in ethanol is inadequate to preserve the degraded, low molecular-weight DNA inside apoptotic cells: this portion of DNA leaks out during subsequent rinsing and staining procedures, and therefore less DNA in these cells stains with any DNA fluorochrome (21). The appearance of cells with low DNA stainability, lower than that of G_1 cells ('sub-G_1 peaks', 'A_0' cells) in cultures treated with various cytotoxic agents, has been considered to be a marker of cell death by apoptosis (31–33).

The degree of DNA degradation varies depending on the stage of apoptosis, cell type and often the nature of the apoptosis-inducing agent. The extractability of DNA during the staining procedure (and thus separation of apoptotic from live cells) also varies. It has been noted that addition of phosphate–citric acid buffer (PC) at pH 7.8 to the rinsing fluid enhances extraction of the degraded DNA (34, 35). This approach can be used to control the extent of DNA extraction from apoptotic cells to the desired level, to obtain their optimal separation by flow cytometry.

Since the cell cycle position of the non-apoptotic cells can be estimated (*Figure 3*), the method described in *Protocol 6* can be applied to investigate the cell cycle specificity of apoptosis. Another advantage of this method is its simplicity, and applicability to any DNA fluorochrome (33) or instrument.

In addition to apoptotic cells, the 'sub G_1' peak, can also represent mechanically damaged cells, cells with lower DNA content (e.g. in a sample containing cell populations with different DNA indices), or cells with different chromatin structure (e.g. cells undergoing erythroid differentiation) in which the accessibility of DNA to the fluorochrome is diminished (36).

It should be stressed that this method can be used only on cells that are fixed in ethanol or permeabilized with detergents. Cell fixation in formaldehyde or glutaraldehyde crosslinks low molecular-weight DNA to other constituents and precludes its easy extraction. Only very late apoptotic cells, in which DNA degradation is significantly advanced, show diminished DNA content following fixation with crosslinking agents.

Protocol 6. DNA staining following extraction of low molecular-weight DNA

A. *Preparation of buffers and solutions*

1. Mix 192 ml of 0.2 M Na_2HPO_4 with 8 ml of 0.1 M citric acid; pH of this solution (PC buffer) should be 7.8.
2. Dissolve 200 μg of PI in 10 ml of HBSS.
3. Add 50 Kunitz units of DNase-free RNase A.
4. Boil RNase for 5 min if RNase is not DNase-free.
5. Prepare fresh staining solution before each use.

B. *Method*

1. Fix cells in suspension in 70% ethanol by adding 1 ml of cells suspended in HBSS (10^6–5×10^6 cells) into 9 ml of 70% ethanol in a tube on ice.
   - Cells can be stored in fixative at −20°C for several weeks
2. Centrifuge cells, decant ethanol, suspend cells in 10 ml of HBSS and centrifuge for 5 min at 100 *g*.
3. Suspend cells in 1 ml of HBSS, into which 0.2–1.0 ml of the PC buffer may be added
   - Add less PC buffer (e.g. 0–0.2 ml) if DNA degradation in apoptotic cells is extensive and DNA is easily extractable.
   - Add more of the buffer (up to 1.0 ml) if DNA is not markedly degraded and there are problems separating apoptotic cells from G_1 cells due to their overlap on DNA content frequency histograms.
4. Incubate at room temperature for 5 min.
5. Centrifuge cells for 5 min at 100 *g* and add 1 ml of HBSS containing PI and RNase A.
6. Incubate cells for 30 min at room temperature.
7. Analyse cells by flow cytometry.
   (a) Use laser excitation at 488 nm or blue light (BG12 filter) and measure red fluorescence (>600 nm) and forward light scatter.
   (b) Apoptotic cells should have a diminished forward light scatter signal and decreased PI fluorescence compared to the cells in the main peak (G_1).
   (c) Cellular DNA may be stained with other fluorochromes instead of PI, and other cell constituents may also be counterstained

Protocol 7. Simultaneous staining of DNA and protein with DAPI and sulforhodamine 101, respectively

1. After step no. 3 of *Protocol 6*, suspend the cell pellet in 1 ml of a staining solution that contains the following:
   - Triton X100, 0.1% (v/v)
   - $MgCl_2$, 2 mM
   - NaCl, 0.1 M
   - PIPES buffer, 10 mM; final pH 6.8
   - DAPI, 1 μg/ml
   - Sulforhodamine 101 (Molecular Probes, Inc.), 20 μg/ml (sulforhodamine is unnecessary if only DNA content is measured)
2. Analyse cells by flow cytometry.
   (a) Use excitation with UV light (e.g. 351 nm argon ion line, or UG1 filter for mercury lamp illumination)
   (b) Measure the blue fluorescence of DAPI in a band from 460–500 nm
   (c) Measure the red fluorescence of sulforhodamine 101 at >600 nm

Figure 4 illustrates the results obtained by the procedure described in *Protocol 6*. As is evident, apoptotic cells are well separated from live cells based on differences in DNA content. They also have diminished protein content. Note that S phase cells in the control population have higher protein content than the apoptotic cells. As mentioned, the degree of separation may be modified by varying the concentration of the PC buffer. This buffer can also be added to the staining solution (e.g. in 1:10 to 1:5 proportion) to extract low molecular-weight DNA from apoptotic cells, when unfixed cells are stained with DNA specific dyes following cell permeabilization with detergents.

11. Denaturability of DNA *in situ*

One of the early events of apoptosis is chromatin condensation. The sensitivity of DNA *in situ* to denaturation is much higher in condensed chromatin, e.g. as in mitotic or G_0 cells (24). Chromatin condensation in apoptotic cells is accompanied by even higher DNA denaturability (37).

A flow cytometric method for measurement of DNA *in situ* sensitivity to denaturation, developed in our laboratory (24), is based on the metachromatic property of the dye Acridine Orange (AO). This dye, under certain conditions, can differentially stain double-stranded (ds) versus single-stranded (ss) nucleic acids. Namely, AO intercalates into dsDNA and, in this mode of binding, upon excitation with blue light emits green fluorescence. In contrast, when AO interacts with ss nucleic acids, it results in red fluorescence. In this method (*Protocol 8*), the cells are briefly pre-fixed in formaldehyde followed

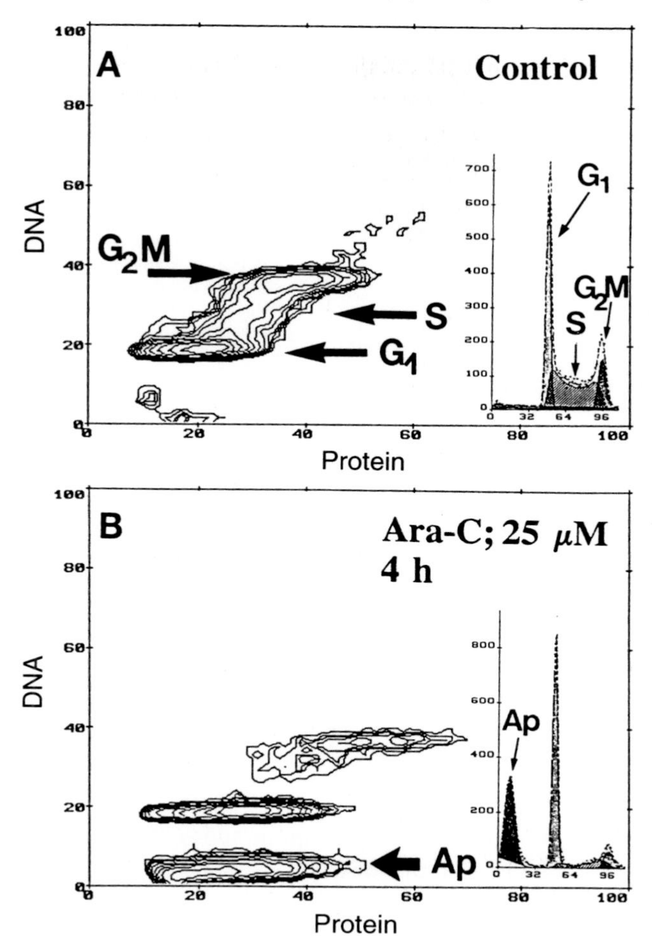

Figure 4. Bivariate analysis of DNA and protein content distributions of (A) control HL-60 cells and (B) cells treated with 25 μM 1β-arabinofuranosylcytosine (Ara-C) for 4 h. DNA content frequency histogram of these cells is shown in the inset. The presence of apoptotic cells (Ap) in Ara-C treated culture correlates with disappearance of S phase cells. A clear distinction between Ap and non-apoptotic cells was obtained by rinsing the ethanol-prefixed cells with PC buffer (pH 7.8), as described in the procedure.

by ethanol. RNA is removed from the cells by preincubation with RNase A and DNA is then denatured *in situ* by brief exposure to 0.1 M HCl. The cells are then stained with AO at pH 2.6; the low pH of the staining reaction prevents DNA renaturation (24). Apoptotic cells, in which the DNA is more sensitive to denaturation, have more intense red and reduced green fluorescence, compared to non-apoptotic cells, which have intense green and minimal red fluorescence (*Figure 5*; ref. 37).

Protocol 8. DNA *in situ* sensitivity to denaturation

Reagents

- HBSS solution containing Mg^{2+} but no phenol red
- Phosphate-buffered saline (PBS) containing 1 mM $MgCl_2$ may be used instead of HBSS
- 1% methanol-free formaldehyde in PBS, pH 7.4
- 70% ethanol
- 0.1 M HCl
- AO: dissolve 1 mg of AO (chromatographically tested; Molecular Probes, Inc.) in 1 ml of distilled water (stock AO solution; stable when stored at 4°C and in the dark for several months)
- AO staining solution: mix 90 ml of 0.1 M citric acid with 10 ml of 0.2 M Na_2HPO_4 (final pH is 2.6); add 0.6 ml of the AO stock solution (1 mg/ml) to 100 ml of this buffer; the final AO concentration is 6 μg/ml (this solution is stable for several weeks when stored in the dark, at 4°C)
- RNase: 1 mg of RNase A in 1 ml of distilled water (use DNase-free RNase). Less pure preparations require 2–3 min heating at 100°C to inactivate DNase

Method

1. Rinse cells once with HBSS, suspend in HBSS (10^6–10^7 cells/ml).
2. Fix cells by transferring 1 ml of the above cell suspension into a tube containing 9 ml of 1% formaldehyde in PBS, on ice. After 15 min centrifuge.
3. Suspend cell pellet in 5 ml of PBS, centrifuge.
4. Suspend cell pellet in 1 ml of PBS, transfer the suspension into a tube containing 9 ml of 70% ethanol, on ice.
   - Samples should be kept in ethanol for a minimum of 4 h.
   - The cells may be transported or stored for several weeks while suspended in ethanol.
5. Centrifuge the fixed cells for 5 min at 100 *g*. Suspend the cell pellet (10^6–10^7 cells) in 1 ml of HBSS.
6. Add 0.2 ml of RNase A solution. Incubate at 37°C for 1 h.
7. Centrifuge for 5 min at 100 *g* and resuspend cells in 1 ml of HBSS.
8. Withdraw a 0.2 ml aliquot of cell suspension in HBSS and transfer it to a small (e.g. 5 ml volume) tube.
9. Add 0.5 ml of 0.1 M HCl, at room temperature (do not use cold solutions).
10. After 30 sec add 2.0 ml of AO solution, also at room temperature.
11. Transfer this suspension to the flow cytometer and measure cell fluorescence. The fluorescence pattern remains stable for several hours.
    - AO is excited with blue light; you may use the 457 or 488 nm laser lines
    - the green fluorescence is measured at 530±20 nm
    - red fluorescence is measured at >640 nm

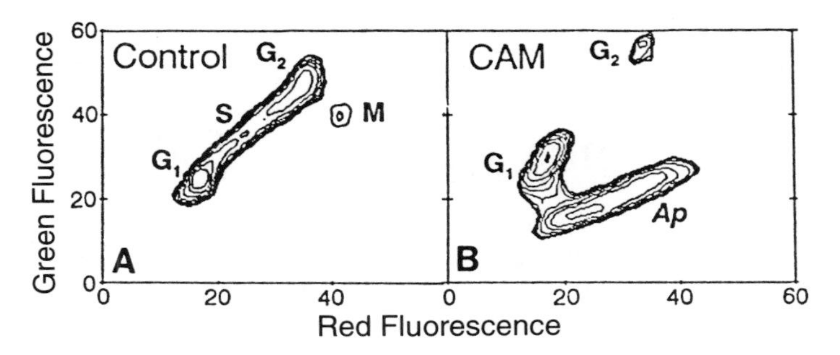

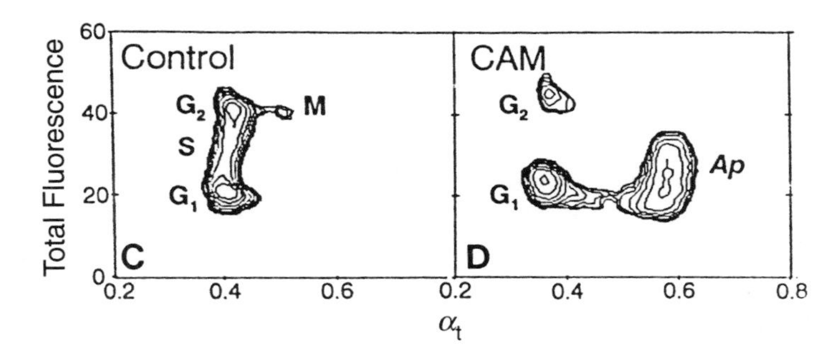

Figure 5. Detection of apoptotic cells based on differences in sensitivity of DNA *in situ* to denaturation. HL-60 cells, untreated (A,C) or treated with the 0.15 μM CAM (DNA topoisomerase I inhibitor) for 4 h (B,D) were stained with AO as described in *Protocol 8*. The data shown in panels A and B represent cellular green and red fluorescence (integrated areas of the pulses), whereas panels C and D represent the same cell population as in A and B, but plotted with respect to total (red plus green) cellular fluorescence versus α_t, where α_t is the ratio of red to total fluorescence (24). Under these staining conditions the cellular green and red fluorescence intensities are proportional to the content of double- and single-stranded DNA, respectively, whereas total fluorescence corresponds to total DNA content (and thus correlates with the cell position in the cell cycle) and α_t reflects the portion of denatured DNA (37, 38). In the untreated cultures only mitotic cells have high red fluorescence and correspondingly, a high α_t value. Camptothecin selectively triggers apoptosis of S phase cells; note their increased red fluorescence (B) and high α_t value (D).

The critical aspect of the procedure described in *Protocol 8* is the choice of the proper AO concentration. This may vary depending on the instrument, due to dye diffusion from the sample to the sheath flow in the measuring channel. In general, if the red fluorescence is too low (and thus apoptotic cells are poorly discriminated), as may occur in instruments that have long channels or in some sorters, this can be compensated for by increasing the AO concentration in the staining solution. A detailed description of the strategies for optimal detection of cells with condensed chromatin using AO, controls, standards, discrimination of mitotic and G_0 cells, is presented elsewhere (1, 38).

12. Labelling of DNA strand breaks in apoptotic cells

As a consequence of activation of the apoptosis-associated endonuclease, apoptotic cells are characterized by massive numbers of DNA strand breaks. Their presence can be detected by labelling the 3′-OH termini in DNA breaks with biotin or digoxygenin-conjugated nucleotides, in a reaction catalysed by exogenous TdT (22, 23) or DNA polymerase (23, 39). In this procedure (*Protocol 9*) the cells are pre-fixed in formaldehyde, which, unlike fixation in ethanol, prevents the extraction of degraded, low molecular-weight DNA. Thus, the DNA content of early apoptotic cells, and with it the number of DNA strand breaks, are not markedly diminished compared to cells fixed with ethanol alone.

The detection of DNA strand breaks *in situ* provides a rather specific assay of apoptosis by flow cytometry. Necrotic cells, or cells with primary breaks induced by X-ray irradiation (up to a dose of 25 Gy), have significantly lower incorporation of b-dUTP compared to apoptotic cells (22). However, additional studies are needed to show whether necrosis induced by other factors can be distinguished from apoptosis by this method. The method is applicable to clinical material and was used to measure apoptosis of blast cells in peripheral blood or bone marrow induced during chemotherapy of leukaemias (40) or in solid tumours, from needle biopsy specimens (41).

12.1 DNA strand labelling with biotinylated dUTP (b-dUTP)

Protocol 9. Labelling of DNA strand breaks with biotinylated dUTP

Reagents

- 1st fixative: 1% methanol-free formaldehyde (paraformaldehyde; Polysciences Inc.) in PBS, pH 7.4
- 2nd fixative: 70% ethanol
- TdT reaction buffer[a] (5 × concentrated; reagent no. 3): 1 M potassium (or sodium) cacodylate, 125 mM Tris–HCl, pH 6.6 at 4°C, bovine serum albumin (BSA), 1.25 mg/ml, 10 mM cobalt chloride
- TdT storage buffer[a] (reagent no. 4): 0.2 M potassium cacodylate, 1 mM EDTA, 4 mM 2-mercaptoethanol, 50% (v/v) glycerol, pH 6.6 at 4°C, TdT, 25 units per 1 μl, biotin-16-dUTP (b-dUTP), 50 nmol in 50 μl
- Saline–citrate buffer: 4× concentrated solution of saline–sodium citrate buffer (SSC) (0.6 M NaCl, 0.06 M Na citrate) (dilute 20× concentrated solution fivefold with distilled water), fluorescein-conjugated avidin, 2.5 μg/ml (final concentration), Triton X100, 0.1% (v/v), BSA, 10 mg/ml
- Rinsing buffer: HBSS containing Triton X100, 1% (v/v), BSA, 5 mg/ml
- Staining buffer: HBSS containing PI, 5 μg/ml, DNase-free RNase A, 1 mg/ml

Method

1. Fix cells in suspension in 1% formaldehyde for 15 min on ice.
2. Sediment cells, resuspend pellet in 5 ml HBSS, centrifuge.
3. Resuspend cells in 70% ethanol on ice. The cells can be stored in ethanol, at −20°C for up to three weeks.
4. Rinse cells in HBSS and resuspend the pellet (less than 10^6 cells) in 50 μl of a solution which contains:
   - 10 μl of the reaction buffer
   - 1 μl (1 μg) of biotin-16-dUTP
   - 0.2 μl (5 units) of TdT in storage buffer
   - 38.8 μl distilled H_2O
5. Incubate cells in this solution for 30 min at 37°C.
6. Add 1.3 ml of the rinsing buffer and centrifuge for 5 min at 100 *g*.
7. Resuspend the pellet in 100 μl of the saline–citrate buffer containing fluoresceinated avidin.
8. Incubate at room temperature for 30 min.
9. Add 1.3 ml of the rinsing buffer and centrifuge as above.
10. Rinse again with rinsing buffer.
11. Resuspend the cell pellet in 1 ml of PI/RNase A solution.
12. Incubate for 30 min at room temperature.
13. Analyse cells by flow cytometry.
    (a) Illuminate with blue (488 nm) light
    (b) Measure fluoresceinated avidin green fluorescence (530±30 nm)
    (c) Measure PI red fluorescence (>620 nm)

[a] The reaction and storage buffers as well as fluorescein-labelled avidin can be obtained from Boehringer Mannheim; cat. nos. 220 582 and 1093 070.

12.2 DNA strand break labelling with digoxygenin-conjugated dUTP

An alternative to b-dUTP is the use of digoxygenin conjugated dUTP (d-dUTP), which is incorporated into DNA strand breaks by TdT similarly as b-dUTP, but instead of avidin, is detected by fluoresceinated antibody to digoxygenin. ONCOR, Inc. provides a kit (ApoptTag^{TM} kit, cat. no. S7100-kit) to identify apoptotic cells, which utilizes d-dUTP and TdT. A description of the method is included with the kit. The results obtained with this kit compare favourably with the data obtained with b-dUTP (42).

As shown in *Figure 6*, simultaneous detection of DNA strand breaks and

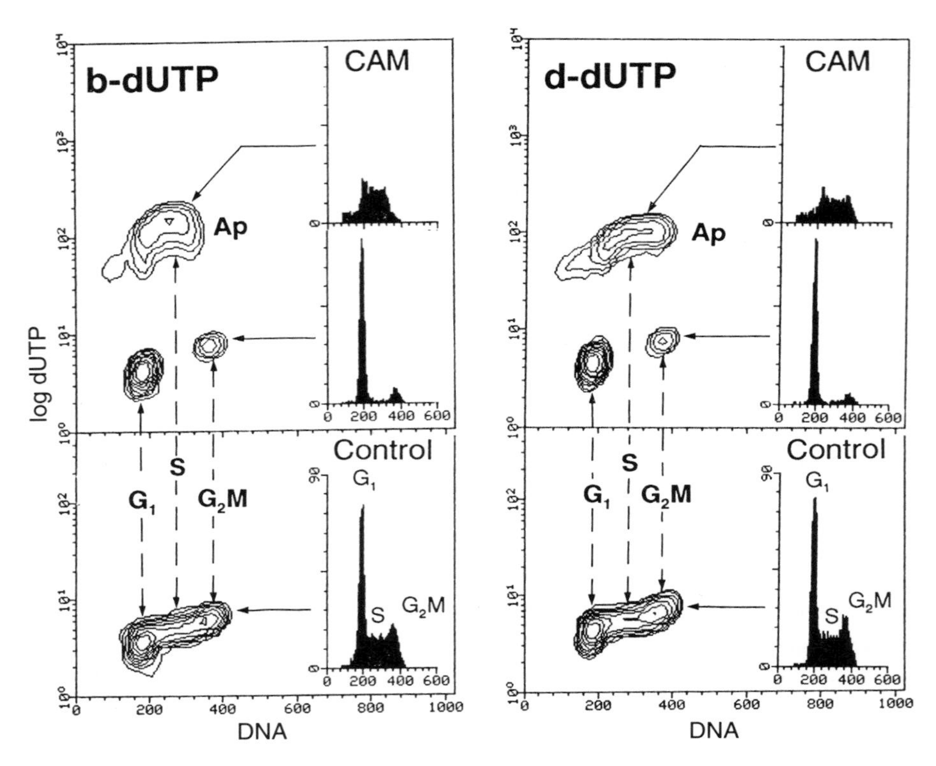

Figure 6. Detection of the apoptosis-associated DNA strand breaks in HL-60 cells treated with the DNA topoisomerase I inhibitor, camptothecin. DNA strand breaks in apoptotic cells (Ap) are labelled with biotinylated dUTP (b-dUP) or digoxygenin labelled dUTP (d-dUTP) in a reaction catalysed by terminal deoxynucleotidyl transferase (TdT). Simultaneous counterstaining of DNA by PI allows correlation of the presence of DNA strand breaks with cell position in the cell cycle. By gating analysis of the dUTP labelled and unlabelled cell populations, the cell cycle distribution of non-apoptotic and apoptotic cells, respectively, can be estimated (see inserts). In contrast to the staining of DNA in ethanol fixed cells (*Figure 3*), only a minor amount of DNA is extracted from apoptotic cells by this method. Note the S phase specificity of apoptosis after treatment with camptothecin.

analysis of DNA content, which is provided by the method described in *Protocol 9*, makes it possible to identify the cell cycle position of cells in both apoptotic and non-apoptotic populations.

13. Which method to choose?

The applicability of each of the methods presented in this chapter varies, depending on the cell system, nature of the inducer of cell death, mode of cell death, the particular information that is being sought (e.g. specificity of apoptosis with respect to the cell cycle phase or DNA ploidy), and the technical restrictions (e.g. the need for sample transportation, type of flow cytometer available, etc.).

Generally, the methods based on analysis of the integrity of the plasma membrane (exclusion of PI, FDA hydrolysis, cell sensitivity to trypsin and DNase I), although simple, and of low cost, fail to identify early apoptotic cells. They can be used to identify necrotic cells, cells damaged mechanically, or in the late stages of apoptosis.

Identification of apoptotic cells is more complex. First, there is a growing body of evidence that in many cell systems apoptosis may be 'atypical', lacking one or more of the features (e.g. internucleosomal DNA degradation, nuclear fragmentation, etc.) that characterize classical apoptosis such as that induced by corticosteroid hormones in thymocytes (43). In other situations, some features associated with apoptosis, such as internucleosomal DNA degradation, may accompany cell necrosis (9). Clearly, the use of a technique that is based on the detection of such a feature will fail to identify the 'atypical' apoptotic or necrotic cell.

Even when apoptosis is typical, none of the described methods when used alone, can provide total assurance of its detection. For practical purposes the most specific approach appears to be the method based on labelling DNA strand breaks with biotin or digoxygenin-conjugated nucleotides. The number of DNA strand breaks in apoptotic cells is so large, that the labelling of the latter cells in this assay appears to be intense enough to discriminate them from necrotic cells. The labelling of cells with primary DNA strand breaks induced by high doses of radiation is also much lower compared to apoptotic cells (22, 40). The assays of DNA strand breaks, however, are complex and rather expensive but were found to be sensitive and convenient for detecting apoptosis in clinical samples (22, 40, 41). Methods based on the extractability of low molecular-weight DNA from apoptotic cells followed by DNA content measurement, are simple and low cost.

Apoptosis can be identified with a higher degree of assurance by using more than one viability assay on the same sample. This may include simultaneous assessment of plasma membrane integrity (e.g. exclusion of PI, hydrolysis of FDA), the function of some organelles such as mitochondria (Rh123 uptake) or lysosomes (AO uptake), chromatin condensation (DNA denaturability), and/or DNA cleavage. For example, preservation of plasma membrane integrity combined with chromatin condensation and extensive DNA breakage will be a more specific marker of apoptosis than each of these features alone.

Regardless of the assay used to identify apoptosis, the mode of cell death should be positively identified by inspection of cells under light or electron microscopy. Morphological changes during apoptosis have a very specific pattern and should be deciding in situations when there is ambiguity regarding the mechanism of cell death.

Acknowledgements

This work was supported by USPHS Grant CA 28704. Dr Jianping Gong, on leave from the Department of Surgery, Tongji Hospital, Wuhan, China, was supported by the fellowship from the 'This Close' for Cancer Research Foundation. Dr Shinsuke Hara, on leave from the School of Medicine, University of Nagasaki, Japan, was supported by a fellowship from his University.

References

1. Darzynkiewicz, Z., Li, X., and Gong, J. In *Methods in cell biology* (ed. Z. Darzynkiewicz, H. A. Crissman, and J. P. Robinson), Academic Press, San Diego (in press).
2. Wyllie, A. H. (1992). *Cancer Metastasis Rev.*, **11**, 95.
3. Arends, M. J., Morris, R. G., and Wyllie, A. H. (1990). *Am. J. Pathol.*, **136**, 593.
4. Compton, M. M. (1992). *Cancer Metastasis Rev.*, **11**, 105.
5. Tomei, L. D. and Cope, F. O. (ed.) (1991). In *Apoptosis: the molecular basis of cell death*. Cold Spring Harbor Laboratory Press, Plainview, NY.
6. Chang, W. P. and Little, J. B. (1992). *Int. J. Radiat. Biol.*, **60**, 483.
7. Radford, J. R. (1991). *Int. J. Radiat Biol.*, **59**, 1353.
8. Catchpoole D. R. and Stewart, B. D. (1993). *Cancer Res.*, **53**, 4287.
9. Collins, R. J., Harmon, B. V., Gobe, G. C., and Kerr, J. F. R. (1992). *Int. J. Radiat. Biol.*, **61**, 451.
10. Meyaard, L., Otto, S. A., Jonker, R. R., Mijnster, M. J., Keet, R. P. M., and Miedema, F. (1992). *Science*, **257**, 217.
11. Hickmann, J. A. (1992). *Cancer Metastasis Rev.*, **11**, 121.
12. Pezzela, F., Tse, A. G. D., Pulford, K. A. F., Gatter, K. C., and Mason, D. Y. (1990). *Am. J. Pathol.*, **137**, 225.
13. Potten, C. P. (1992). *Cancer Metastasis Rev.*, **11**, 179.
14. Schwartzman R. A. and Cidlowski, J. A. (1993). *Endocr. Rev.*, **14**, 133.
15. Horan, P. K. and Kappler, J. W. (1977). *J. Immunol. Methods*, **18**, 309.
16. Hamori, E., Arndt-Jovin, D. J., Grimwade, B. G., and Jovin, T. M. (1980). *Cytometry*, **1**, 132.
17. Darzynkiewicz, Z., Williamson, B., Carswell, E. A., and Old, L. J. (1984). *Cancer Res.*, **44**, 83.
18. Helson, L., Traganos, F., and Miller, D. (1981). *Cancer Clin. Trials*, **4**, 415.
19. Johnson, L. U., Walsh, M. L., and Chen, L. B. (1980). *Proc. Natl. Acad. Sci. USA*, **77**, 990.
20. Darzynkiewicz, Z., Traganos, F., Staiano-Coico, L., Kapuscinski, J., and Melamed, M. R. (1982). *Cancer Res.*, **42**, 799.
21. Darzynkiewicz, Z., Bruno, S., Del Bino, G., Gorczyca, W., Hotz, M. A., Lassota, P., and Traganos, F. (1992). *Cytometry*, **13**, 795.
22. Gorczyca, W., Bruno, S., Darzynkiewicz, R. J., Gong, J., and Darzynkiewicz, Z. (1992). *Int. J. Oncol.*, **1**, 639.
23. Gorczyca, W., Gong, J., and Darzynkiewicz, Z. (1993). *Cancer Res.*, **53**, 1945.

24. Darzynkiewicz, Z., Traganos, F., Carter, S., and Higgins, P. J. (1987). *Exp. Cell Res.*, **172**, 168.
25. Dive, C., Gregory, C. D., Phipps, D. J., Evans, D. L., Milner, A. E., and Wyllie, A. H. (1992). *Biochim. Biophys. Acta*, **1133**, 275.
26. Ormerod, M. G., Collins, M. K. L., Rodriguez-Tarduchy, G., and Robertson, D. (1992). *J. Immunol. Methods*, **153**, 57.
27. Swat, W., Ignatowicz, L., and Kisielow, P. (1991). *J. Immunol. Methods*, **137**, 79.
28. Stohr, M. and Vogt-Schaden, M. (1980). In *Flow cytometry IV* (ed. O. D. Laerum, T. Lindmo, and E. Thorud), pp. 96–9. Universitetforslaget, Bergen.
29. Wallen, C. A., Higashikubu, R., and Roti Roti, J. L. (1983). *Cell Tissue Kinet.*, **16**, 357.
30. Pollack, A. and Ciancio, G. (1990). In *Flow cytometry* (eds. Z. Darzynkiewicz and H. A. Crissman), pp. 19–24. Academic Press, San Diego, CA.
31. Umansky, S. R., Korol', B. R., and Nelipovich, P. A. (1981). *Biochim. Biophys. Acta*, **655**, 281.
32. Nicoletti, I., Migliorati, G., Pagliacci, M. C., Grignani, F., and Riccardi, C. (1991). *J. Immunol. Methods*, **139**, 271.
33. Telford, W. G., King, L. E., and Fraker, P. J. (1991). *Cell Prolif.*, **24**, 447.
34. Gorczyca, W., Gong, J., Ardelt, B., Traganos, F., and Darzynkiewicz, Z. (1993). *Cancer Res.*, **53**, 3186.
35. Gong, J., Li, X., and Darzynkiewicz, Z. (1993). *J. Cell Physiol.*, **157**, 263.
36. Darzynkiewicz, Z., Traganos, F., Kapuscinski, J., Staiano-Coico, L., and Melamed, M. R. (1984). *Cytometry*, **5**, 355.
37. Hotz, M. A., Traganos, F., and Darzynkiewicz, Z. (1992). *Exp. Cell Res.*, **201**, 184.
38. Darzynkiewicz, Z. (1994). In *Methods in Cell Biology* (ed. Z. Darzynkiewicz, H. A. Crissman, and J. P. Robinson) Academic Press, San Diego, 15–38.
39. Gold, R., Schmied, M., Rothe, G., Zischler, H., Breitschopt, H., Wekerle, H., and Lassman, H. (1993). *J. Histochem. Cytochem.*, **41**, 1023.
40. Gorczyca, W., Bigman, K., Mittelman, A., Ahmed, T., Gong, J., Melamed, M. R., and Darzynkiewicz, Z. (1993). *Leukemia*, **7**, 659.
41. Gorczyca, W., Tuziak, T., Kram, A., Melamed, M. R., and Darzynkiewicz, Z. (1994). *Cytometry*, 15, 169.
42. Li, X., James, W. M., Traganos, F., and Darzynkiewicz, Z. (1994). *Biotech. Histochem.* (in press)
43. Cohen, G. M., Sun, X.-M., Snowden, R. T., Dinsdale, D., and Skilleter, D. N. (1992). *Biochem. J.*, **286**, 331.
44. Del Bino, G., Lassota, P., and Darzynkiewicz, Z. (1991). *Exp. Cell Res.*, **193**, 27.

9

Assays of cell growth and cytotoxicity

PHILIP SKEHAN

1. Introduction

This chapter describes non-radioactive assays commonly used to measure changes in cell growth and cytotoxicity induced in a cell population by treatment with agents such as drugs, hormones, nutrients, and irradiation. It focuses specifically on mammalian cell cultures, but the methods can be used with a wide variety of cells, tissues, and microorganisms. Because of the ease and speed with which large numbers of samples can be processed by optical plate readers, most of the protocols which follow are designed specifically for 96-well plates, but are easily adaptable to other culture vessels.

2. Growth and cytotoxicity assays

There are four types of non-radioactive cell growth and cytotoxicity assays: (i) cell or colony counts, and assays of (ii) macromolecular dye binding, (iii) metabolic impairment, and (iv) membrane integrity. No single method is universally appropriate for all situations. Each has limitations, and all are subject to potentially serious artefacts under certain circumstances.

The ideal assay is simple, rapid, reliable, sensitive, quantitative, objective, inexpensive, and foolproof. Cell and colony counts are time consuming, tedious, and sensitive to minor variations in methodology. Cell counts enumerate morphologically intact cells but do not distinguish between living and dead cells. Colony counts often require subjective judgements about what constitutes a colony, and are subject to a wide variety of troublesome artefacts that greatly complicate their interpretation.

Dye-binding assays probably come closest to fulfilling the ideal requirements for growth and cytotoxicity assays. They are simple, rapid, reliable, sensitive, and quantitative, but do require access to a 96-well plate reader or spectrometer. They also fail to distinguish between living and dead cells *per se*, although with proper experimental design they will distinguish between net cell killing, net growth inhibition, and net growth stimulation.

Metabolic impairment assays measure the decay of enzyme activity or metabolite concentration following toxic insult. They are generally more complex and artefact-prone than dye-binding assays. Their validity requires that the same precise conditions be met that are required by a bimolecular chemical reaction analysis. Deviation from these conditions can lead to extremely serious errors that invalidate the assay. Metabolic impairment assays are nevertheless popular because they distinguish between normal and reduced levels of cellular metabolism, which is a surrogate index of metabolic viability though not necessarily an accurate predictor of proliferative capability.

Membrane integrity assays measure the ability of cells to exclude impermeant extracellular molecules. They can be either colorimetric or fluorescent, and require the same instrumentation as dye-binding assays. They tend to be less artefact-prone than metabolic impairment assays, but have the same ability to estimate 'viability', which in this case is the ability to distinguish between the normal and impaired exclusion of extracellular molecules. Membrane integrity assays are complicated by the fact that living cells slowly accumulate probe molecules; their protocols must be very carefully optimized. Like metabolic impairment assays, they provide a surrogate index of viability that is not always an accurate predictor of proliferative capacity.

3. Experimental design

Growth is the change in a quantity over time (1). It is a rate process that requires at least two separate measurements. In growth end-point assays, measurements are made at the start and end of an experiment. In kinetic assays, intermediate measurements are also recorded to produce a time series of values.

Five different types of samples are required to determine the biological effect of an experimental treatment upon a cell population: (i) medium blanks—growth medium with no cells or drugs, (ii) drug blanks—growth medium with drug but no cells; (iii) time zero values—cells plus medium at the start of an experiment, and end of assay values for (iv) control cells plus medium, and (v) test cells plus medium.

The population size of samples is calculated by subtracting medium blanks from time zero and control measurements, and drug blanks from test samples receiving drugs (1). This produces three final population measurements: the time zero population size (Z), the control population size (C) at the end of the assay, and the test population size (T) at the end of the assay. The net effect of a treatment is determined in the following way (1):

- Net growth stimulation has occurred if T is greater than C.
- Net cell killing has occurred if T is less than Z.
- Net growth inhibition has occurred if T is greater than Z but less than C.

4. Cell cultures

4.1 Seeding density

Seeding density depends on cell size, growth rate, and assay duration. It must be determined individually for each cell type (2). In a 2–3 day assay, seeding densities are typically in the range of 5–25 thousand cells per well (kcpw) in 96-well microtitre plates. Time zero values must be sufficiently above background to allow net cell killing to be accurately quantitated, whereas end of assay values must remain within the linear range of the assay, typically 1.5–2.0 absorbance units.

4.2 Dissociation and recovery

Chemical and enzymatic dissociation are traumatic processes that overtly kill some cells while sensitizing survivors to other toxic insults. This sensitization can produce false–positive artefacts with cells subsequently exposed to growth inhibitors or cytotoxins. Dissociated cells require a recovery period before an experiment is started. One day is usually sufficient (2).

4.3 Drug solubilization

Stock solutions of polar compounds are made in water, buffer, or medium then diluted in complete growth medium to the final test concentration (2). Non-polar compounds are dissolved in a solvent such as dimethylsulphoxide (DMSO) or ethanol (ETOH) and filter sterilized (0.22 μm pores). A 1:1 mixture of DMSO and ETOH is also a good solvent and it evaporates more slowly than ETOH and chemically sterilizes most test materials. DMSO is toxic to cells at concentrations above 0.1–1.0% Preliminary experiments should be carried out to determine its toxicity threshold for each individual cell line. Ethanol is usually growth stimulatory in the 1–2% range.

4.4 Assay duration

Assay duration is determined by two factors: (i) the length of time cells need to respond to an experimental treatment; and (ii) the length of time that cells can grow before nutrient depletion sets in. Depletion typically develops within 3–4 days after plating unless cultures are re-fed (3). Once it begins, a progressive deterioration of cellular health and viability develops rapidly, and becomes a major artefact in data interpretation. Depletion can be calibrated by comparing the day by day growth kinetics of cultures that receive no feeding to cultures that are fed daily (3). The two curves begin to diverge when depletion sets in. This normally sets the upper limit to assay duration if cultures are not fed. If experiments must be continued for longer periods, then medium must be replaced. Above densities of near-confluency, most types of cells require daily feeding. With cytotoxicity assays, a 36–48 h assay period following a 1 day recovery period is usually adequate to detect the effect of a drug while avoiding the need to re-feed in mid-experiment.

4.5 Control wells on every plate

Every test plate requires its own control wells. Because temperature and humidity are not uniform in tissue culture incubators, plates located at different positions tend to experience different local micro-environments which differentially affect cell growth. Control values from one plate are often different from and cannot reliably be used for test samples on other plates.

5. Dye-binding assays

This section describes assays for culture protein, biomass, double-stranded polynucleic acid, and DNA. The protein and biomass assays are colorimetric, and measure the binding of coloured dyes to cell cultures fixed with trichloroacetic acid (TCA). The polynucleic acid and DNA assays are fluorescent, and measure the amount of fluorochrome intercalation into RNA or DNA in freeze-thaw permeabilized cells.

5.1 Optimizing and validating dye-binding assays

Dye-binding assays are bimolecular chemical reactions that use a change in the amount of one reactant (the dye) to measure the amount of a second (the cellular target). Valid binding assays must meet several requirements which serve collectively to guarantee that a single cell in one test sample produces the same signal as an identical cell in another. If these conditions are not met, then two samples cannot be quantitatively compared. The necessary conditions are established by optimizing a protocol for each experimental system in each laboratory using the cells and reagents of that laboratory. They should not be adopted in a cookbook manner from the literature. The dye-binding assays described in this chapter have the desirable characteristic that their optimized protocols, once established, are identical for a wide variety of cell types (2, 4, 5). This is not true with most other assays.

5.1.1 Supramaximal dye concentration

The concentration of a dye must be sufficiently high (supramaximal) to saturate all target binding sites. A supramaximal concentration is any dye concentration sufficiently high that its further increase causes no further increase in staining intensity (4, 5). If the condition is not met, dye-binding does not accurately measure the amount of a macromolecular target.

5.1.2 Supramaximal dye volume

To saturate target binding sites, dye concentration and volume of staining solution must both be supramaximal for assay validity (5). A supramaximal volume is one on the plateau region of the curve relating staining volume to measured signal for a concentration of target molecule higher than will ever be encountered experimentally.

5.1.3 Supramaximal staining time

A third requirement for the validity of dye-binding assays is that the staining time must be supramaximal (4, 5). A supramaximal staining time is one in which the signal being measured has reached a stable plateau value that does not increase further with time. It should be determined using a culture density higher than will ever be encountered experimentally.

5.1.4 Washing away unbound dye

Several of the assays require that unbound dye be washed away with an appropriate buffer at the end of a staining period. Washing can cause the loss of cell-associated dye. Rinse procedures must be carefully optimized to determine the minimum number of rinses necessary to remove excess unbound dye without desorbing any significant amount of cell-associated stain (4, 5). Rinses should be performed rapidly and rinse solutions removed from cultures as quickly as possible; never leave them sitting in wells. Failure to reach a stable plateau invalidates the assay.

5.2 Protein and biomass stains

Many colorimetric biological stains bind electrostatically to macromolecular fixed ions of opposite electrical charge. The binding and desorption of these dyes can be controlled by varying pH. Dilute acids and bases, such as 1% acetic acid and 10 mM unbuffered Tris base, can be used to bind and solubilize the dyes (1).

Under mildly acidic conditions, anionic dyes bind to basic amino residues, serving as cellular protein stains. They can be extracted and solubilized by weak base. Protein stains suitable for growth and cytotoxicity assays (*Table 1*) include sulforhodamine B (SRB), Orange G (ORG), Bromophenol Blue (BPB), and chromotrope 2R (CTR).

Under mildly basic conditions, cationic dyes bind to negative fixed charges of proteins, RNA, DNA, and glycosaminoglycans, serving as stains for culture biomass. They can be quantitatively extracted from cells and solubilized for absorbance measurement by weak acid. Biomass stains suitable for growth and cytotoxicity assays (*Table 1*) include Thionin (TNN), Azure A (AZRA), and Toluidine Blue O (TBO). Phenosafranin and Safranin O are also acceptable as biomass stains.

5.2.1 Cell fixation

Protein and biomass are both measured on fixed cell cultures. Trichloroacetic acid (TCA) and perchloric acid (PCA) are the fixatives of choice (1). They fix cells instantly, permeabilize their membranes to allow rapid dye penetration,

Table 1. Optimal staining protocols for selected dyes

	Protein stains [a]			Biomass stains [b]		
	ORG	**BPB**	**CTR**	**TNN**	**AZRA**	**TBO**
Concentration (% w/v)	1.5	1.0	0.25	0.3	0.25	0.2
Staining solution	AcOH	AcOH	AcOH	Tris	Tris	Tris
Minimum stain time (min)	5	20	30	10	30	10
Number of washes	3	4	3	4	4	3
Solubilizing solution [c]	Tris	Tris	Tris	AcOH	AcOH	AcOH
Optimal wavelength (nm)	478	590	508	594	628	626
Resolution (kcpw [d])	2–3	3–4	4–5	>5	>10	>10

[a] Protein stains: ORG (Orange G), BPB (Bromophenyl Blue), CTR (chromotrope 2R).
[b] Biomass stains: TNN (thionin), AZRA (Azure A), TBO (Toluidine Blue O).
[c] Solubilizing solutions: 1% acetic acid (AcOH); 10 mM unbuffered Tris base.
[d] kcpw = thousands of cells per well.

extract small-molecular-weight metabolites, greatly strengthen cell adhesion, and harden cells to withstand rough handling. Formaldehyde is less effective as a fixative. Glutaraldehyde, because of its bifunctionality, can potentially interfere with the assay, and should not be used. Organic fixatives such as ethanol and methanol cause significant cell lysis, and are not appropriate for these assays.

Protocol 1 describes a method for the *in situ* TCA fixation of adherent cell cultures (1). Once fixed, cells are extremely resistant to damage and dislodgement. Solutions can be poured directly on to fixed plates from a beaker or even a running water tap with no risk of cell dislodgement. The only delicate part of the fixation process is the initial addition of the fixative; it must be added gently without producing any fluid shearing forces that would dislodge cells before they become fixed.

Protocol 1. *In situ* TCA fixation of adherent cells in 96-well microtitre plates (1, 4)

1. Grow cells with 200 μl of medium per well. To fix, gently layer 50 μl of cold 50% TCA on top of the 200 μl already in the well to produce a final TCA concentration of 10%.
2. Incubate plates for 30 min at 4°C to complete fixation.
3. Wash plates five times with distilled or tap water.
4. Air dry at room temperature; dried plates can be stored indefinitely before further processing.

For suspensions of single cells and very small aggregates (about five cells or less), the modified fixation procedure of *Protocol 2* should be used. For larger aggregates, which are not attached to the plastic by this fixation method, the propidium iodide assay of *Protocol 4* should be used.

Protocol 2. *In situ* TCA fixation of single cell suspensions in 96-well microtitre plates (4)

1. Grow cells with 200 μl of medium per well. Lay plates on a stable benchtop and allow cells to settle to the bottom of the wells. Gently layer 50 μl of cold 80% TCA on top of the 200 μl already in the well to produce a final TCA concentration of 16%; cells must be in contact with tissue culture plastic at the instant that the TCA front reaches them in order to be attached.
2. Leave the plates undisturbed on the benchtop for 30 min to complete the fixation process.
3. Wash five times with distilled water as with adherent cultures (*Protocol 1*).
4. Air dry at room temperature; dried plates can be stored indefinitely before further processing.

5.2.2 Sulforhodamine B (SRB) colorimetric assay of cell protein

Sulforhodamine B (SRB) is a bright pink aminoxanthene dye with two sulphonic groups. It is one of a family of related dyes, such as Naphthol Yellow S and Coomassie Brilliant Blue, that are used widely as protein stains in histochemistry and molecular biology. Under mildly acidic conditions, SRB binds to basic amino acid residues of TCA-fixed proteins. It provides a sensitive index of culture cell protein that is linear with cell number over a cell density range of more than 100-fold. Of more than 20 protein and biomass stains investigated for possible use in microtitre assays, SRB provided the best combination of staining intensity and signal-to-noise ratio (4). It provides a stable end-point that does not have to be measured within any fixed period of time. Once stained and air dried, plates can be kept indefinitely without deterioration before solubilization and reading. SRB has proven particularly useful in very large-scale drug screening (2).

Protocol 3. Sulforhodamine B (SRB) assay for culture cell protein in 96-well microtitre plates (4)

1. Harvest plates by fixing with TCA as described in *Protocol 1* or *2*.
2. Stain TCA-fixed cells for at least 30 min with 0.4% (w/v) SRB in 1% acetic acid.

Protocol 3. *Continued*

3. Remove SRB and quickly wash plates four times with 1% acetic acid to remove unbound dye.
   - Wash quickly to avoid desorbing bound dye molecules; pour acetic acid on to plates from a large beaker; remove by turning plate upside down and flicking to shake out residual fluid.
   - Do not aspirate individual wells; this takes too much time and will desorb protein-bound dye molecules, causing artefacts in the data.
   - Complete all four washes of one plate before going on to the next.
4. Air dry until no standing moisture is visible (over night is recommended); plates can be stored indefinitely before proceeding.
5. Solubilize bound SRB by adding a fixed volume (usually 100–200 μl) of 10 mM unbuffered Tris base (pH 10.5) to each well and shaking for at least 5 min on a shaker platform.
6. Read absorbance in a 96-well plate reader; the optimal wavelength is 564 nm.

Note: The assay is very sensitive; confluent cultures often give absorbances that are above the linear range of the assay (about 1.8 absorbance units). It is frequently desirable to use a suboptimal wavelength so that higher cell densities will remain within the range of linearity. Wavelengths in the 490–530 nm range work well for this purpose.

5.2.3 Other protein stains

Orange G, Bromophenol Blue, and chromotrope 2R have chemistries identical to SRB, and stain TCA-fixed protein nearly as well (4). Their optimized assays follow *Protocol 3* modified as shown in *Table 1*. Coomassie Brilliant Blue and Naphthol Yellow S are much less satisfactory.

5.2.4 Cellular biomass assays

Thionin, Azure A, and Toluidine Blue O are biomass stains that approach the protein stains in sensitivity (4). Protocols for their use are identical to those for the protein stains, except that the acidic and basic solutions are reversed. Use the modifications of *Protocol 3* indicated in *Table 1*.

5.2.5 Propidium iodide fluorescent assay of double-stranded polynucleic acid

Propidium iodide (PI) is a general biomass stain that binds to RNA, DNA, protein, and glycosaminoglycans (GAGs). Its fluorescence increases by as much as 100-fold when it intercalates into double-stranded sequences of RNA and DNA (6). This allows double-stranded polynucleic acids to be measured in the presence of free, unbound PI provided there are no significant amounts of proteins or GAGs present. An important exception is a narrow excitation peak centred at 531 nm with a band width of approximately 2 nm (5). At this excitation wavelength, fluorescence is independent of cell protein over a broad range of emission wavelengths centred at 604 nm. Using narrow-bandwidth 531/604 nm excitation/emission filters, double-stranded polynucleic acid content (RNA + DNA) of cultures can be measured in the presence of complete growth medium and serum, thus providing the basis for an exceptionally simple assay that requires no washing steps at all (5). Cultures are permeabilized by freeze–thawing, then an aliquot of propidium iodide is added to each well, and after a short incubation period PI fluorescence is read in a fluorescent plate reader at wavelengths of 530/590 nm or 530/620 nm (*Protocol 4*).

These wavelength combinations reflect RNA more than DNA. Because RNA usually changes earlier than protein in cellular growth shifts, the PI assay tends to detect cellular responses earlier than protein, biomass, and DNA assays. It is usually the method of choice where short assays are desirable. It is also the method of choice for extreme assay simplicity.

The excitation filters of fluorescent plate readers are broad band, and do allow some fluorescence excitation of protein-bound PI. This produces a low level of background fluorescence from serum in the growth medium. Sensitivity of the PI plate reader assay is 1000–3000 cells per well in 5% fetal calf serum, comparable to the SRB assay. With 10% serum, resolution is reduced by a factor of 2. These values are obtained with ordinary 96-well plates; somewhat better resolution can be achieved with more expensive low-fluorescence plates.

The PI assay was originally designed for use with human lymphoma cells, which aggregate extensively and cannot be measured with most other assays (5). It works equally well with single cell suspensions, adherent cultures, and bacteria; it is probably the simplest of all plate reader-based growth and cytotoxicity assays.

With all of the cell lines that we have examined, a single freeze–thaw cycle has been sufficient to permeabilize cells to PI, even though it does not always kill cells. Detergent-permeabilization must not be used with this assay; detergents have complex and unpredictable effects, and can lead either to strong quenching or high background levels of fluorescence. High salt, EDTA, and proteinase K treatment do not significantly influence the sensitivity of the PI assay (5).

Unlike many fluorescent assays, PI fluorescence is not sensitive to pH under the conditions of *Protocol 4*. Thus ordinary media like RPMI can be used without worrying about the pH shifts caused by cell growth or plate removal from a CO_2 incubator. Fluorescence is steady for several hours, although polynucleic acids sometimes degrade over longer periods of time.

Protocol 4. Propidium iodide (PI) assay of cellular RNA and DNA in 96-well plates (5)

1. Harvest plates by freezing at −30°C for at least 2 h; frozen plates can be stored indefinitely.
2. Thaw at 50°C for 15 min.
3. Add 50 μl/well of a 200 μg/ml PI stock[a] in distilled water to the 200 μl of growth medium in each well; the final PI concentration is 40 μg/ml.
4. Incubate in the dark at room temperature for 60 min.
5. Read fluorescence at 531/604 nm excitation/emission wavelengths in a spectrofluorimeter or at 530/590–620 nm in a fluorescent plate reader.
   - Excitation wavelength is critical, emission is not.
   - In a spectrofluorimeter, readings are independent of culture protein and serum to at least 10% fetal calf serum; in a plate reader there is a slight background fluorescence from serum and cell protein that cuts sensitivity in half when serum is raised from 5 to 10%.

[a] Wrap PI solutions in foil and protect from light.

5.2.6 Hoechst 33258 fluorescence assay of DNA

Hoechst 33258 is a UV-excited blue bisbenzimidazole dye which selectively intercalates into A-T rich regions of DNA, undergoing a fluorescence enhancement in the process. Unlike PI, the Hoechst dye is specific for DNA rather than for macromolecules or polynucleic acids generally. It provides a fluorescence signal that is linearly proportional to DNA content over a wide range of DNA values. The Hoechst 33258 assay is nearly as simple as the PI assay, the only additional step being an initial removal of medium from culture wells (7). This step is required to allow the addition of an EDTA-high NaCl buffer which partly dissociates DNA, allowing better dye access and therefore brighter fluorescence. The assay is comparable to the PI assay in sensitivity. It is well suited for use with firmly adherent cultures, but is less suitable for single cell suspensions or adherent lines that produce viable subpopulations of floating cells.

Protocol 5. Hoechst 33258 fluorescence assay of cellular DNA in 96-well microtitre plates (7)

Materials

- Hoechst 33258 solutions wrapped in foil and protected from light
- TNE buffer: 10 mM Tris, 1 mM EDTA, 2 M NaCl, pH 7.4

Method

1. Remove medium from plates.
2. Freeze plates at −80°C; store frozen until ready for further processing.
3. Thaw plates.
4. Add 100 μl distilled water to each well and incubate for 1 h at room temperature.
5. Refreeze at −80°C for 90 min.
6. Thaw to room temperature.
7. Add 100 μl of TNE containing 20 μg/ml of Hoechst 33258 dye; after dilution, final Hoechst 33258 concentration is 10 μg/ml.
8. Incubate in dark for 90 min.
9. Read in fluorescent plate reader at 350/460 nm excitation/emission wavelengths.

6. Metabolic impairment assays

6.1 Neutral Red viability assay

Neutral Red is a vital dye that accumulates in the lysosomes of living, uninjured cells. Dead and severely traumatized cells lose their ability to accumulate and retain Neutral Red, providing the basis for a simple colorimetric viability assay used widely in cellular toxicology (8).

With vital dye assays, great care must be taken in selecting the length of the dye incubation period. The rate of dye uptake differs widely from one cell type to another, and the uptake itself is sometimes preceded by a lag period of unpredictable length. The 3 h dye incubation period listed in step 5 of *Protocol 6* is only a rough guide; the staining duration should be individually optimized for each cell line. Because the dye is simultaneously taken up and lost by cells, the staining time must be sufficiently long to allow intracellular Neutral Red levels to reach a stable equilibrium value. This usually requires a few tens of minutes. Cells lose Neutral Red rapidly once the dye is removed. It is advisable to conduct efflux studies to determine just how rapidly the loss occurs, and then to ensure that the wash period is sufficiently short to avoid leakage artefacts.

Protocol 6. Neutral Red cell viability assay (8)

1. Prepare a 0.4% stock solution of Neutral Red in distilled water. This may be stored in the dark at 4°C for several months.
2. Within 24 hours of use, dilute the Neutral Red stock solution 1:80 in growth medium to produce a final concentration of 50 μg/ml; optimize this concentration individually for each cell line.

Protocol 6. *Continued*

3. Prewarm the Neutral Red solution to 37°C, and centrifuge at 700 *g* for 5 min immediately before use to remove undissolved dye crystals.
4. Remove medium from the test plates and replace with 0.2 ml of Neutral Red solution per well.
5. Incubate for ~3 h at 37°C; individually optimize incubation time for each cell line.
6. Remove the Neutral Red solution, and rinse the wells once with 0.2 ml per well of 4% formaldehyde containing 1% calcium chloride. This washes away residual Neutral Red and fixes the cells to prevent their detachment during dye solubilization. The rinsing process must be completed within 1–2 min to avoid loss of intracellular dye.
7. Add 0.2 ml of solubilization fluid (1 ml glacial acetic acid in 100 ml of 50% ethanol) to each well.
8. Solubilize the dye for 15 min on a microtitre plate shaker.
9. Read absorbance at 540 nm.

Note: Wrap Neutral Red solutions in foil and protect from light

6.2 MTT cell viability assay

Tetrazolium reactions have been used widely to histolocalize dehydrogenase enzyme activity. The tetrazoliums do not react with the dehydrogenases *per se*, but rather with their reaction products NADH and NADPH. The reaction reduces water-soluble tetrazolium salts to a highly coloured and insoluble formazan which precipitates out of solution in the immediate vicinity of the reaction. A yellow-coloured tetrazolium, 3-(4,5-dimethylthiazole-2-yl)-2,5-diphenyl tetrazolium bromide (MTT), is converted by the reduction reaction into a purple formazan, and has been adapted for use in microtitre plates as a cellular viability assay (9).

The MTT assay has the same validity requirements as a Lineweaver–Burke analysis: (i) initial reaction velocity must be measured over (ii) a linear range with (iii) a supramaximal concentration of MTT, and (iv) a protocol that is insensitive to microenvironmental fluctuations in parameters such as pH and temperature. The assay only measures cell number accurately when the ability of cells to reduce MTT is rigidly constitutive; if the ability is non-constitutive, differences in microenvironment or past histories of two otherwise identical cells will produce different levels of enzyme expression, invalidating their quantitative comparison.

The MTT assay has a number of serious problems (10). The intensity of MTT reduction varies considerably from one cell line to another, and frequently declines with increasing culture age. The reaction tends to be non-linear with cell number, and both the reaction rate and the spectral absorbance curve vary with pH. The MTT reaction rate varies with cellular and medium glucose levels; it is only linear for short periods of time, and the period of linearity varies with the cell line employed. There are frequently long lag periods between the addition of MTT and the onset of its metabolism, making it virtually impossible to measure initial reaction velocity accurately. The supramaximal concentration of MTT varies widely with cell line.

These several difficulties collectively can make apparent IC_{50}s (drug concentration producing 50% growth inhibition) vary by as much as 20-fold depending on MTT concentration, culture age, population density, and length of assay. In using the MTT reaction as a viability assay, great care must be exercised to optimize it fully for each individual cell line and to work strictly within its limits of documented validity.

The MTT assay described in *Protocol 7* is an unusually well-optimized method that is distinctive in its use of a pH 10.5 buffer that increases sensitivity and reduces or eliminates many of the artefacts normally associated with the assay (11).

Protocol 7. MTT cell viability assay in 96-well microtitre plates (11).

1. Prepare a 5× stock (~2 μg/ml) of MTT in phosphate-buffered saline (PBS). The final concentration is typically about 0.4 μg/ml, but should be individually optimized for each cell line.
2. Add 50 μl of the MTT stock to the 200 μl of medium in each well.
3. Incubate cultures at 37 °C for a period that is typically 0.5–4 h in length but must be individually determined for each cell line.
4. Aspirate off medium from each well. For non-adherent cells, leave about 20 μl of medium in the wells to avoid loss of cells, then centrifuge plates at 200 *g* for 5 min.
5. Add 25 μl of Sorenson's glycine buffer.[a]
6. Add 200 μl of DMSO to each well and incubate for 10 min while vibrating on a plate shaker to solubilize the formazan crystals.
7. Read absorbance at 570 nm. This wavelength is critical, do not change.

[a] *Sorenson's glycine buffer*: 0.1 M glycine, 0.1 M NaCl, adjusted to pH 10.5 with 0.1 M NaOH.

7. Membrane integrity assays

7.1 Fluorescein diacetate (FDA)

Fluorescein diacetate (FDA) is an electrically neutral non-fluorescent molecule. Viable cells accumulate FDA intracellularly and hydrolyse it to an anionic reaction product, fluorescein, which is highly fluorescent. Fluorescein is retained intracellularly by living cells for short periods of time, causing them to become temporarily fluorescent. Dead cells lack the ability to accumulate and hydrolyse FDA, and are therefore non-fluorescent. Both FDA and fluorescein are non-toxic; they provide the basis for a vital assay of cell metabolic viability that allows test cultures to be subsequently reused for other purposes (12).

Because FDA is hydrolysed by serum, producing high levels of background fluorescence, assays must be conducted in serum-free solution. Fluorescein is pH-sensitive, and undergoes a fluorescence enhancement with increasing alkalinity. The pH must be kept rigidly constant by using a well-buffered solution such as PBS as a solvent and rinsing solution. Serum-free non-CO_2-buffered media such as PDRG (13) or GIBCO's CO_2-independent growth medium are also suitable, and are more physiological than simple buffers such as PBS. Alternatively, serum-free RPMI and MEM (minimal essential medium) can be used if they are supplemented with a CO_2-independent buffer such as 25 mM Hepes or β-glycerophosphate.

Like all viability assays, the FDA to fluorescein conversion does not, strictly speaking, measure viability *per se*. The conversion is a function of cell volume, cell physiology, and level of hydrolytic enzyme activity as well. However, the assay correlates well with and is linearly proportional to cell number under a variety of conditions (14).

Protocol 8. Fluorescein diacetate cell viability assay in 96-well microtitre plates (14)

1. Prepare a 10 μg/ml stock solution of FDA in DMSO; store at −20°C.
2. Collect test plates; centrifuge at 200 *g* for 5 min.
3. Remove medium by inverting plates and flicking gently.
4. Wash once with PBS or a serum-free, pH stable, non-CO_2-buffered medium.
5. Add 200 μl of prewarmed FDA (10 μg/ml) in the same solution used in step 4 (optimize the exact FDA concentration for each individual cell line).
6. Incubate at 37°C for 60 min; optimize the exact incubation time for each individual cell line with particular attention to a possible lag phase in fluorescence production.
7. Centrifuge plates at 200 *g* for 5 min; remove solutions as in step 3.
8. Add 200 μl of prewarmed PBS or serum-free medium to each well.
9. Read fluorescence immediately at 485/538 nm excitation/emission wavelengths.

Wrap FDA solutions in foil and protect from light

8. Electrical conductivity assays

Living cells condition their growth medium by secreting into it a variety of metabolites that alter the medium's electrical conductivity. Over a period of several hours, medium conductivity comes into a dynamic equilibrium with cell number. From this point on, medium conductivity is linearly proportional to cell number with a wide variety of cell lines and micro-organisms. Because the metabolites which produce this effect are labile, the conductivity change is reversed when cells die. Thus, medium conductivity can be used to monitor both cell growth and cytotoxicity.

CellStat Technologies has developed an automated system that kinetically monitors the medium conductivity changes which attend cell growth and cytotoxicity. The CellStat System^{TM} consists of a special plate lid, a measuring board, a wheel-mounted movable CO_2 incubator, and a DOS 486 computer (*Figure 1*). The plate lid is a printed circuit board that replaces the lid on a 24-well plate. The lid contains an array of electrode pairs that extend into each well. Each electrode pair measures the electrical resistance of the medium in a single well. Resistance measurements are conducted via an I/O interface to an analog measuring board that processes the signals from each well and transmits them to a data acquisition board in the computer. The

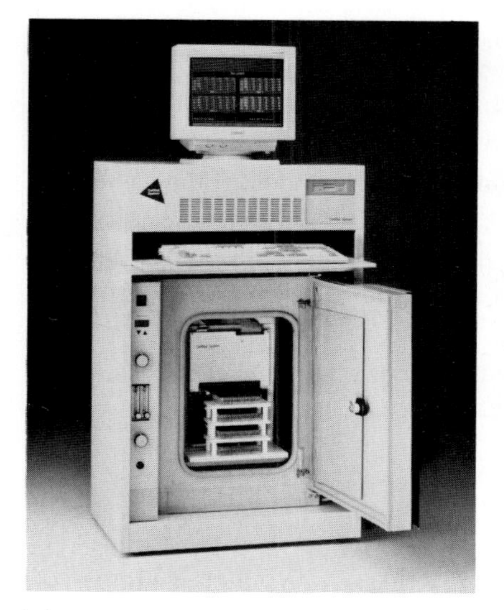

(a)

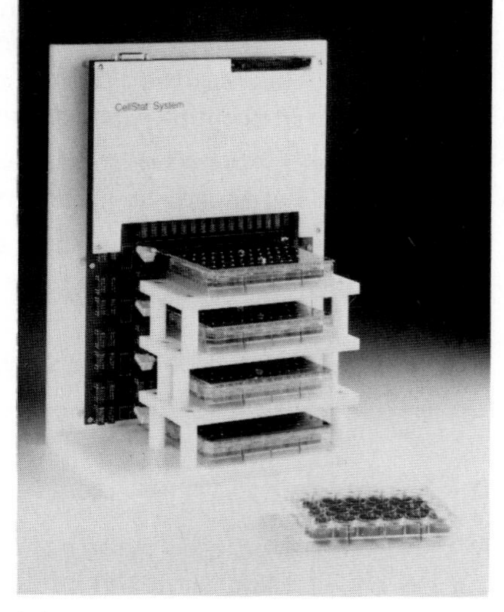

(b)

Figure 1. The CellStat System for measuring medium conductivity: the entire system (a) and the measuring board containing four 24-well plates (b). Cells change the electrical conductivity of the medium as they grow. Changes in conductivity serve as an index of metabolic viability, permitting cell growth and cytotoxicity to be automatically measured as frequently as every 15 min for periods of up to two weeks.

computer calculates conductivity, graphs data, and provides screen and hard-copy reports of both the raw data and summary calculations.

The CellStat System can accept up to four 24-well plates at a time, permitting as many as 96 wells to be monitored simultaneously. Experiments are started by a single keystroke, and continue unattended for up to 2 weeks. Experiments are automatically terminated at a predetermined time, and can be manually stopped at any time. During an experiment, accumulated data can be graphed on screen, and the initial and most recent measurements can be displayed in tabular form. Because of its ability to report data while an experiment is in progress, the CellStat System is particularly useful when brief assays are desired but where the necessary length of an assay cannot be predicted in advance.

Protocol 9. CellStat measurement of cell growth by changes in medium electrical conductivity

1. Seed cells in 24-well plates containing 2 ml of growth medium or test solution per well. Seeding density varies with cell line and assay duration, but is typically in the range of 1×10^4–1×10^5 cells/well. Plates should contain medium blanks, drug blanks, controls, and test samples.
2. Sterilize the CellStat plate lid by soaking in ethanol for 1–2 h. Remove the lid, shake off the excess ethanol, place on to a sterile 24-well plate bottom, and thoroughly dry under sterile conditions. Drying is facilitated by leaning the lid/plate combination at an angle and draining on to an absorbent paper towel.
3. Remove the plate lid from a test plate and replace it with the CellStat electrode array lid; be careful not to bend the electrodes.
4. Enter information about the experiment into the computer. When finished, provide the keystroke that starts the experiment. No further attendance is required.
5. Kinetic data can be displayed while the experiment is in progress.

9. Survivorship assays

In cultures treated with cytotoxic drugs, some cells die and lyse, others die but do not lyse, and still others are mortally injured but do not die or lyse within the time limit of the experiment. Survivors often go through an extended period of proliferative quiescence as part of their response to environmental stresses, and their subsequent outgrowth is often a very slow process requiring weeks or months.

The short assays described in *Protocols 3–9* commonly fail to detect small subpopulations of surviving cells, and do not identify populations of momentarily surviving cells that are destined to die later as the result of experimental insult. Survivorship assays provide a means for identifying such populations. They represent an extended second stage assay that is conducted following completion of a short-term primary assay such as those described in *Protocols 3–9*.

There are two classes of survivorship assays: (i) recovery, and (ii) colony formation. Recovery assays follow bulk cultures for extended periods of time to determine whether they eventually exhibit cellular recovery or delayed mortality. Colony-forming assays convert populations in to single cell suspensions, then determine the proportion of cells that give rise to progressively growing colonies. Colony forming efficiency (CFE) is the percentage of single cells plated that gives rise to multicellular colonies. CFE can be measured by plating cells on a solid substratum such as tissue culture plastic, or by seeding them in a non-adherent medium such as soft agar.

9.1 Adherent versus non-adherent cells

Mammalian cells of solid tissue origin usually require some degree of adhesive contact to other cells or a physical surface. When deprived of these contacts, their proliferation slows and their death rate increases, causing serious artefacts in survivorship assays that restrict adhesion and cell-to-cell contact (15, 16). Soft agar CFE assays are best suited to non-adherent cells that grow naturally as single cell suspensions. Adherent cells tend to exhibit elevated mortality rates in soft agar, a tendency that can be potentiated by cytotoxic insult. The adhesion of cells to a plastic substratum at least partly satisfies their anchorage requirement, and supports better proliferation and viability than growth in soft agar. Long-term recovery and plastic CFE are generally preferable to soft agar assays with adherent cell populations. Recovery assays are in turn preferable to plastic CFE assays because they do not require cytotoxic chemical or enzymatic dissociation, and better maintain cell-to-cell contacts during the highly stressful period of early recovery.

9.2 Long-term recovery (LTR) assay

The simplest survivorship assay is the long-term recovery (LTR), which is based on the fact that living cells secrete metabolic acids that reduce the pH of a culture's growth medium (17). In media containing Phenol Red as the pH dye indicator (most media), this metabolism is evidenced by a change in colour from red to either orange or yellow.

Before starting an LTR assay, perform a primary assay using the methods of *Protocols 3–9* to construct a dose–response curve for a drug of interest. From this, select three drug concentrations which produce effects that are: (a) half-maximal, (b) just maximal, and (c) supramaximal.

Conduct the LTR assay in T25 flasks rather than 96-well plates. Seed cells in 5–10 ml of medium, but otherwise use the same protocol (cells/cm^2, assay duration, etc.) that was employed in the primary assay. At the end of the drug incubation period, replace the test solution with fresh drug-free medium, equilibrate the atmosphere in the flask with that of the incubator chamber, close the flask top tightly, and incubate at 37°C for as long as desired. *Protocol 10* uses 60 days as a standard recovery period, but the assay can be abbreviated or extended indefinitely. Replicate flasks can be collected at intermediate times to explore the possibility that delayed cell killing has occurred.

To use the LTR assay qualitatively, examine the colour of flasks visually 2–3 times weekly. Use cell-free flasks treated with drugs then incubated in drug-free medium as a reference. The survivorship of metabolically viable cells is established when test cultures become less red and more orange or yellow than the reference flasks. Verify microscopically that the survivorship is by the experimental cell population rather than by contaminating micro-organisms. Data are expressed as growth delay – the number of days following drug removal at which cultures are judged to have clearly exhibited metabolic recovery.

To determine whether metabolically recovered cells are also proliferatively viable, dissociate and replate cells. Collect a time zero sample, then incubate replicate flasks for several days. At the end of a normal culture growth period, count cells or conduct one of the assays in *Protocols 3–5*. An increase above time zero value establishes progressive growth and proliferative viability.

Metabolic recovery can be documented quantitatively by measuring the change in culture pH. The absorbance maximum for phenol red in cell-free growth medium in the pH 7.4–7.6 range is approximately 560 nm. As the medium shifts from red toward orange or yellow, the absorbance at 560 nm decreases. If a CO_2-independent medium is used (Section 7.1), pH is stable in air so that aliquots can be collected and read at 560 nm in a microtitre plate reader or spectrophotometer. With CO_2-dependent media, prepare a set of tightly capped reference flasks containing bicarbonate-free medium adjusted to various pH values; use these to estimate visually the pH of experimental flasks.

Protocol 10. Long-term recovery assay for T25 flasks (17)

1. Perform a short-term dose–response analysis in 96-well microtitre plates (*Protocols 3–9*).
2. Repeat the experiment in T25 flasks exactly as in microtitre plates except:
   (a) Plate cells in 5–10 ml of growth medium.
   (b) Scale up inoculum size to correct for different vessel sizes (i.e. use the same number of cells per cm^2 in the T25 flasks that you used in the microtitre plates).

3. Treat flasks with three drugs determined by data from step 1: use concentrations that are (a) half maximal, (b) just maximal, and (c) supramaximal.
4. At the end of the drug incubation period, remove test solution and replace with fresh drug-free medium; if there are a significant number of floating cells in the medium, centrifuge them, resuspend, and place back in the flask together with fresh medium.
5. Equilibrate the atmosphere in the flask for 1–2 h, then cap tightly and incubate at 37 °C.
6. Examine 2–3 times weekly for pH change; when obvious pH change has occurred, evaluate as described in Section 9.2. Growth delays of up to several months can be measured.
7. As a reference, use a cell-free flask treated with drug solution, then incubated during recovery period with fresh drug-free medium.
8. Express recovery as a drug-induced growth delay; quantitate extent of pH change if desired.

9.3 Colony-forming efficiency on tissue culture plastic

Colony formation on a solid substratum such as tissue culture plastic is the preferred method for determining the CFE of adherent cell populations. Cells should be treated experimentally, then dissociated and replated in fresh growth medium. The proper seeding density depends on growth rate and plating efficiency, but typically ranges from a few tens to a few thousands of cells per T25 flask. Seeding density should be selected so that final end-of-assay colony counts are in the range of 100–1000. Increased serum sometimes improves CFE.

The CFE of adherent cells can be determined by counting number of (a) macroscopic colonies visible by eye, (b) microscopic colonies with more than some threshold number (10, 20, or 50) of cells, or (c) microscopic colonies with two or more cells. These methods are not comparable, and commonly give different results. The last criterion is technically the most rigorous, and the first is the least.

Pure single cell suspensions are impossible to prepare for most cell types. Background levels of multicellular aggregates are typically about 5%, but can be as high as 50% in some cases. Aggregates tend to be more viable than single cells, and are more likely to give rise to a progressively growing colony (15). It is critically important to establish the time zero frequency distribution for the number of cells per colony. A colony is any cluster of one or more cells. This distribution becomes the reference to which end-of-assay results are compared. A shift in the distribution toward larger colony sizes is evidence of proliferative recovery. Standard statistical methods can be used to quantitate the shift.

An alternative method is to identify specific colonies at the beginning of an assay and follow them individually throughout the colony-forming process. A series of crosses is made at various points on the bottom of a flask. An optical grid is aligned successively in each of the four quadrants of each cross. The colonies located within the grids are identified and recorded and their size at various times throughout the experiment is estimated. Colony size can be estimated in several ways: (a) number of cells, (b) colony radius or diameter, (c) colony area from photographs or video images, or (d) frequency histogram of cells per colony (it is extremely difficult to count cells accurately in colonies with more than about 20 cells). This approach works well with non-motile cells. If cells and colonies are motile, the same approach can be used but modified by counting colonies per grid without attempting to follow specific individual colonies

Protocol 11. Colony-forming efficiency of adherent cells on tissue culture plastic in T25 flasks

1. Prepare a single cell suspension using *Protocol 1* or *2*.
2. Plate cells in a T25 flask containing 5–10 ml of medium.
   (a) Adjust seeding density so that final end-of-assay colonies are in the range of 100–1000.
   (b) Elevated serum concentrations (10–15%) sometimes improve cell survivorship.
3. Allow cells to attach; the time required varies with the cell line, and can range from a few minutes to overnight; 0.5–3 h is satisfactory for many cell lines.
4. Change the medium to remove unattached cells.
5. Count the time zero number of colonies containing 1, 2, 3, . . . cells per colony (Section 9.3).
6. Incubate cultures until colonies are well developed; this period varies with the cell line, and can range from a few days to several months; change medium if colour begins to turn orange.
7. Count colonies; if cultures are not needed for other purposes, they can be TCA-fixed and stained for further quantitation with SRB (*Protocol 3*) or one of the dyes listed in *Table 1*.

9.4 Colony-forming efficiency in soft agar

CFE assays in soft agar are best suited to cells like those of the haematopoietic system which grow naturally in single cell suspensions. Whereas adherent cells will sometimes grow in soft agar, the environment is extremely nonphysiological and generates a number of complex artefacts that make quantitative analyses dubious (15).

Soft agar assays contain two separate layers: (i) an underlayer which serves as a barrier to prevent cell adhesion to the bottom of the tissue culture chamber, and (ii) an overlayer containing the cells. The underlayer is typically a 0.5% and the overlayer a 0.3% gel. Either agar or agarose can be used. Agar is less pure and contains materials that are growth inhibitory to some cells (15). Methylcellulose, an extremely viscous fluid, can be substituted for agar or agarose in the overlayer.

Protocol 12 describes a survivorship assay in which cells have already been experimentally treated prior to starting the soft agar assay (18). However, the method can also be used as a primary assay in dose–response studies (19). For this purpose, colonies should be grown to observable size before an experimental treatment is begun; their reduction in size or number serves as an index of drug efficacy. The data from soft agar CFE assays are analysed as described in Section 9.3.

Protocol 12. Soft agar colony-forming efficiency in 35 mm dishes

1. Prepare the underlayer.

(a) Make solution A: prepare a double strength solution of growth medium (2× for all components including serum); filter sterilize; divide into two portions; warm to 45°C in water bath.

(b) Make solution B: add 1% (v/v) powdered agar or agarose to distilled water; autoclave 20 min (slow exhaust) to solubilize and sterilize; place in 45°C water bath.

(c) Make solution C (the underlayer): add equal volumes of solutions A and B; mix well; return to 45°C water bath to prevent gelling.

(d) Add 1 ml of solution C to each 35 mm dish; allow solution to gel at room temperature, then incubate dishes at 37°C until further use.

2. Prepare the overlayer.

(a) Prepare solution D: add 0.6% (v/v) powdered agar or agarose to distilled water; autoclave 20 min (slow exhaust) to solubilize and sterilize; place in 45°C water bath.

(b) Prepare solution E: dissociate cells and prepare a 2× single-cell suspension (twice the desired final cell concentration) in double-strength growth medium; equilibrate to 37°C.

(c) Prepare solution F (the cell overlayer): add equal volumes of solutions D and E; mix well without frothing; transfer immediately to a 37°C water bath.

(d) Layer 0.5 ml of solution F on top of underlayer in each 35 mm dish; add immediately before agar begins to gel.

Protocol 12. *Continued*

(e) Leave dishes at room temperature until agar overlayer solidifies.

(f) Add 0.5–1.0 ml of 1× growth medium on top of overlayer.

(g) Transfer dishes to 37°C incubator.

3. Collect time zero plates and construct a time zero frequency histogram of colony size (number of colonies with 1, 2, 3, 4, . . . cells) as described in Section 9.3.

4. Incubate test cultures at 37°C; soft agar assays commonly require 1–3 weeks; to prevent medium depletion, liquid medium should be changed at least twice weekly beginning at the end of the first week.

5. At the end of the assay construct frequency histograms of colony size for test and controls.

6. For counting colonies, contrast can be enhanced by MTT staining (19):

(a) Gently aspirate liquid from top of overlayer.

(b) Add 1 ml of 1 mg/ml MTT made by diluting a sterile-filtered 5 mg/ml MTT stock in PBS 1 to 5 in prewarmed growth medium.

(c) Incubate 4 h.

(d) Remove MTT.

(e) Add 1.5 ml of 2.5% (w/v) protamine sulphate in normal saline solution; incubate at 4°C for 16–24 h, then aspirate free solution; repeat; this procedure is optional, and is designed to increase contrast by reducing background staining of the agar.

(f) Count colonies.

References

1. Skehan, P., Thomas, J., and Friedman, S. J. (1986). *Cell Biol. Toxicol.*, **2**, 357.
2. Monks, A., Scudiero, D., Skehan, P., Shoemaker, R., Paull, K., Vistica, D. T., Hose, C., Langley, J., Cronise, P., Vaigro-Wolff, A., Gray-Goodrich, M., Campbell, H., Mayo, J., and Boyd, M. R. (1991). *J. Natl. Cancer Inst.*, **83**, 757.
3. Skehan, P. and Friedman, S. J. (1984). *Cell Tissue Kinet.*, **17**, 335.
4. Skehan, P., Storeng, R., Scudiero, D., Monks, A., McMahon, J., Vistica, D. T., Warren, J. T., Bokesch, H., Kenney, S., and Boyd, M. R. (1990). *J. Natl. Cancer Inst.*, **82**, 1107.
5. Skehan, P., Bokesch, H., and Williamson, K. (1993). In *Drug resistance on leukemia and lymphoma. The clinical value of laboratory studies* (ed. G.-J. Kaspers), pp. 409–13. Harwood Publishers, Langhorn, PA.
6. Shapiro, H. M. (1988). *Practical flow cytometry*. Alan R. Liss, New York.
7. Rago, R., Mitchen, J., and Wilding, G. (1990). *Anal. Biochem.*, **191**, 31.
8. Borenfreund, E. and Puerner, J. A. (1984). *J. Tissue Culture Methods*, **9**, 7.
9. Mossman, T. (1983). *J. Immunol. Methods*, **65**, 55.

10. Vistica, D. T., Skehan, P., Scudiero, D., Monks, A., Pittman, A., and Boyd, M. R. (1991). *Cancer Res.*, **51**, 2515.
11. Plumb, J. A., Milroy, R., and Kaye, S. B. (1989). *Cancer Res.*, **49**, 4435.
12. Rottman, B. and Papermaster, B. W. (1966). *Proc. Natl. Acad. Sci. USA*, **55**, 134.
13. Vistica, D. T., Scudiero, D., Skehan, P., Monks, A., and Boyd, M. R. (1990). *J. Natl. Cancer Inst.*, **82**, 1055.
14. Nygren, P. and Larsson, R. (1991). *Int. J. Cancer*, **48**, 598.
15. Skehan, P. and Friedman, S. J. (1981). In *The transformed cell* (ed. I. L. Cameron and T. B. Pool), pp. 7–65. Academic Press, New York.
16. Skehan, P. (1991). In *Growth regulation and carcinogenesis* (ed. W. R. Paukovits), Vol. 2, pp. 313–25. CRC Press, Boca Raton.
17. Skehan, P., Williamson, K., and Bowers, G. (1995). *Cell Prolif.* (in press).
18. Freshney, R. I. (1987). *Culture of animal cells. A manual of basic technique*. Alan R. Liss, New York.
19. Alley, M. C., Pacula-Cox, C. M., Hursey, M. L., Rubinstein, L. R., and Boyd, M. R. (1991). *Cancer Res.*, **51**, 1247.

10

Cell synchronization as a basis for investigating control of proliferation in mammalian cells

GARY S. STEIN, JANET L. STEIN, JANE B. LIAN, T. J. LAST, THOMAS OWEN, and LAURA McCABE

1. Introduction

From a historical perspective, synchronized cells have provided the basis for defining regulatory mechanisms operative during specific periods of the cell cycle; for example, those at key transition points that separate G_1, S-phase, G_2 and mitosis as well as those operative when quiescent cells are induced to proliferate. Cell synchrony studies have also contributed to establishing modifications in gene expression which are functionally related to initiation and execution of DNA replication during S-phase and to the onset and completion of mitotic division. More recently this experimental approach has increased our understanding of key modifications in gene expression and the interrelated cellular signalling pathways that transduce, integrate and amplify information that support each stage of the cascade of regulatory events requisite for progression through the cell cycle. This chapter is restricted to methods that have proven to be effective for synchronizing continuously dividing normal diploid and tumour cells as well as for induction of quiescent cells to embark upon a progression of events that leads to cell division. Protocols that can be applied for monitoring cell synchrony will be addressed. In addition, we will evaluate the extent to which cell synchrony can be achieved in a manner that is compatible with physiological regulation that is operative under *in vivo* conditions.

2. Synchronization of continuously dividing cells by thymidine

2.1 General considerations

Synchronization of continuously dividing cells can be effectively achieved by imposing metabolic blocks that meet several criteria. First, cells must be arrested at a specific point in the cell cycle. Second, cells must be permitted to progress through other stages of the cell cycle to reach the arresting point. Third, the block must be rapidly reversible with minimal perturbations of biochemical, cellular or molecular parameters of proliferation. Fourth, phenotypic properties characteristic of specialized cells that are expressed during proliferation should be minimally affected by the cell synchronization protocol.

The use of excess thymidine was the first widely accepted method for inducing cell synchrony and remains one of the most effective techniques (1, 2). By treatment with two sequential 'thymidine blocks' a synchronous population of cells can be obtained at the beginning of S phase. This method can be utilized to synchronize both suspension and monolayer cells and is readily applicable to both normal diploid, transformed, and tumour cells. It should be emphasized that for this method to function optimally all cells must be undergoing exponential growth. A description of the double thymidine block procedure has been reported for tumour cells by Stein and Borun (3) and Holthuis *et al.* (4). An elegant assessment of key parameters associated with thymidine synchronization has also been provided (5–8).

The synchronization protocol is based on imposing the first thymidine block on exponentially growing cells for a minimum period of time equivalent to $G_2 + M + G_1$ but not exceeding 16 h. The rationale is that in inhibiting ribonucleotide reductase activity, thymidine, at high concentrations, inhibits DNA synthesis in S-phase cells by depleting the nucleotide precursor pools of dCTP. Cells in G_2, M and G_1 are not affected by excess thymidine treatment and continue to traverse the cell cycle until reaching the G_1/S phase boundary when the onset of DNA synthesis is inhibited. As a result, at completion of the initial thymidine block approximately half the cell population is uniformly distributed throughout S-phase and the other half is at the beginning of S-phase (see *Figure 1*). The thymidine block is then released for a period of time exactly equivalent to S-phase. This 'release' permits the cells accumulated at the G_1/S-phase boundary and those blocked throughout S-phase to pass through S-phase. Now, with half the cell population in early G_2 and other cells distributed between G_2, M and G_1, a second thymidine block is imposed for a period of time equivalent to $G_2 + M + G_1$, but maximally 16 h. At completion of the second thymidine block all cells are at the beginning of S-phase. Upon release of the second thymidine block the cells synchronously traverse S-phase, G_2, mitosis and G_1.

Experience dictates that thymidine block should generally not exceed 16 h to prevent breakdown of polyribosomes. The concentration of thymidine required to execute this synchronization procedure is dependent upon the cell type. The appropriate thymidine concentration must be experimentally determined by defining the minimal concentration to inhibit DNA synthesis by 95% within 15 min but permit complete resumption of DNA synthesis in all cells immediately following removal of excess thymidine from the culture media. We developed the protocol described in this chapter for synchroniza-

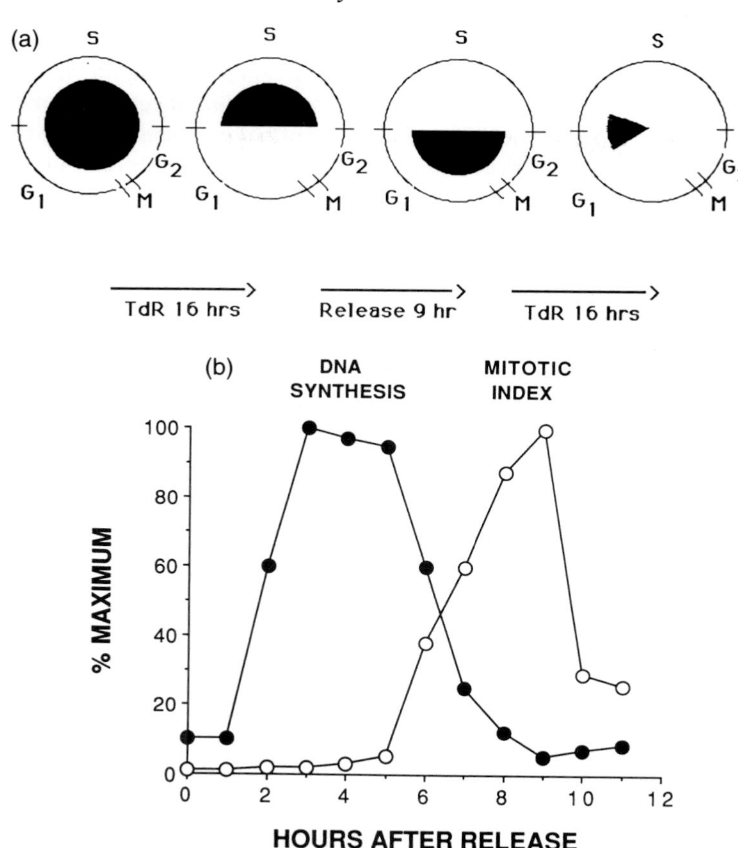

Figure 1. (a) Diagram of double thymidine block method of cell synchrony. (b) DNA synthesis and mitotic index following release from second thymidine block; levels are expressed as a percentage of maximal [^{3}H]thymidine incorporation or as a percentage of the maximal number of mitotic cells counted in a visual field.

tion of exponentially growing HeLa S3 human cervical carcinoma cells in monolayer or suspension culture and primary cultures of normal diploid rat osteoblasts or 17/2.8 rat osteosarcoma cells in monolayer culture on cover slips, glass tissue culture plates or plastic tissue culture dishes. This method is based on a doubling time of 20 h with the following times associated with each stage of the cell cycle: G_1, 6 h; S-phase, 9 h; G_2, 4 h; and mitosis, 1 h.

For the protocols in this chapter, several cell lines and cultures were used. Each of the cell lines were maintained in different medium formulations as described below.

(i) Suspension cultures of HeLa S3 cells are maintained in Joklik-modified Eagle's minimal essential medium (Gibco BRL) and supplemented with

3.5% horse serum (Gibco BRL). Suspension cultures of HeLa S3 cells are maintained in Erlenmeyer flasks or carboys ranging in size from 250 ml to 35 litres. A sterile magnetic stir bar is placed in the bottom of each flask and a sterile cotton plug is inserted in the neck of the flask. The flasks are placed on a magnetic stirrer and maintained at 37°C in a tissue culture incubator or 'warm room'.

(ii) Monolayer cultures of HeLa S3 cells are maintained in Eagle's minimal essential media (Gibco BRL) and supplemented with 3.5% fetal calf (Gibco BRL) and 3.5% horse serum. Monolayer cultures are maintained at 37°C in a tissue culture incubator with a moist, 5% CO_2 atmosphere.

(iii) Primary cultures of normal diploid, calvarial-derived rat osteoblasts are maintained in Eagle's minimal essential medium (Gibco BRL) and supplemented with 10% fetal bovine serum (Gibco BRL).

(iv) ROS 17/2.8 rat osteosarcoma cells are maintained in F12 medium (Gibco BRL) and supplemented with 2% fetal calf and 5% horse serum.

All the cell culture media were prepared by dissolving powdered medium in double glass distilled water and filter sterilizing with a cellulose acetate membrane filter unit (0.2 μm pores) (Corning). Immediately before use, culture medium is supplemented with serum, 2 mM L-glutamine, 1.5% penicillin and 0.5% streptomycin.

Protocol 1. Double thymidine block synchronization of suspension cultures of HeLa S3 cells

Tissue culture and reagents

- Appropriate culture medium
- Thymidine (Sigma): 100 mM (50×) stock solution in serum-free medium and filter sterilize
- Deoxycytidine (Sigma): 24 μM solution in fresh culture medium

Method

1. Dilute exponentially growing cells at a concentration of 5×10^5/ml with fresh medium to a final concentration of 3.5×10^5 cells/ml, and add thymidine to a final concentration of 2 mM.
2. Incubate at 37°C for a mimimum period of time equivalent to G_2 + M + G_1, but not to exceed 16 h.
3. To release the first thymidine block pellet the cells by centrifugation at 600 *g* for 5 min, carefully pour off the growth medium, and wash the cell pellet in 200 volumes of serum-free medium at 37°C to maximally eliminate thymidine-containing medium in contact with the cell pellet.

Resuspend the cells to a final concentration of 3.5×10^5 cells/ml in fresh growth medium containing deoxycytidine.

4. After completion of the 9 h release period, initiate a second thymidine block:
   (a) Dilute the suspension cultures with fresh medium to a final concentration of 3.5×10^5 cells/ml
   (b) Add thymidine to a final concentration of 2 mM.
5. Incubate at 37°C for a period of time equivalent to $G_2 + M + G_1$, but maximally 16 h.
6. To release the cells from the second thymidine block centrifuge at 600 *g* for 5 min as described in step 2 and then resuspend the cells in fresh growth medium containing deoxycytidine to a final concentration of 3.5×10^5 cells/ml.

2.2 Rationale for steps in thymidine synchronization protocol

A 12–16 h treatment of exponentially growing HeLa cells with an S-phase blocking agent during the initial block will allow all cells that were in G_2 mitosis or G_1 at the initiation of the block to progress to the G_1/S-phase boundary, whereas cells undergoing DNA replication will be immediately arrested in S-phase. Thus, at the completion of the first thymidine block, approximately 50% of the cells will accumulate at a G_1/S-phase transition point and the other half will be arrested at various points in S-phase.

The 9 h release period from the first thymidine block permits all cells to exit S-phase. Washing and resuspending the cells in medium containing deoxycytidine facilitates rapid reversal of the thymidine block by compensating for the effect of an expanded thymidine pool. The 50% of the cells that accumulated at the G_1/S-phase boundary during the first thymidine block and at various points during S-phase will be distributed between G_2, mitosis and G_1.

A 12–16 h treatment with excess thymidine during the second block accumulates all cells at the G_1/S-phase boundary. S phase can be followed by determining cellular levels of histone mRNA by Northern blot or slot blot analysis using a ^{32}P-labelled histone gene probe (9).

Although time-consuming, autoradiography following [^{3}H]thymidine labelling of cultures provides the most direct approach to monitor cell synchronization, since visual examination of the autoradiographic preparations unquestionably indicates which cells are replicating DNA. The specific protocols on autoradiography of monolayer cultures have been described comprehensively by Baserga and Malamud (10). S-phase cells can also be visualized by *in situ* hybridization using a ^{35}S-radiolabelled histone gene probe (11). Although requiring specialized instrumentation, cell cycle progression may be monitored by flow cytometry (see Chapter 2).

Protocol 2. Thymidine synchronization of monolayer cultures

Tissue culture and reagents

- As for *Protocol 1*.

Method

1. Initiate a thymidine block procedure in cultures at a cell sensitivity that will permit active growth throughout the time course of the synchronization procedure. Impose the first thymidine block by removing the growth media by aspiration and provide fresh medium containing 2 mM thymidine. Block the cells for 12–16 h.
2. Release the cells from the first block for 9 h by removing the thymidine-containing medium by aspiration and washing the monolayers twice with an equal volume of serum free media (at 37°C) prior to replacement with normal growth medium containing deoxycytidine.
3. Following the 9 h release period, impose a second thymidine block by addition of thymidine to a final concentration of 2 mM from a 50× stock solution in serum-free medium.
4. After 12–16 h release the second thymidine block by removing the thymidine-containing medium by aspiration and washing the monolayers twice with an equal volume of serum free medium (at 37°C) and feeding with normal growth medium containing deoxycytidine.

2.3 Modification of thymidine synchronization for specialized applications

During the past several years variations of the thymidine block procedure have been developed for synchronization of continuously dividing cells. Examples include substitution of aphidicolin (12) (Sigma) (5 μg per ml for HeLa cells, osteosarcoma cells or osteoblasts) or hydroxyurea (Sigma) (1 mM for HeLa cells, osteosarcoma cells or osteoblasts) for thymidine during the second 'S-phase block'.

2.4 Precautions that should be exercised during thymidine synchronization

To maximize the extent to which synchrony is achieved and to minimize introduction of reagent-related artefacts, the following precautions are important.

(i) It is essential to maintain the cells at 37°C throughout the synchronization procedure. Even a slight decrease in temperature will result in significant delays in cell cycle progression. Optimal conditions for cell synchronization are achieved by carrying out the entire procedure in a 37°C environmental room using an ambient temperature centrifuge for harvesting suspension culture cells.

(ii) Minimal concentrations of trypsin and/or EDTA should be used for subculturing monolayer cells prior to synchronization. Cell viability will be increased and exponential growth will resume with minimal delay.

(iii) When releasing thymidine-blocked suspension cells by centrifugation, the cell pellet should be adequately drained to maximize removal of thymidine-containing media.

(iv) When adding or removing media from monolayer cultures during synchronization steps, caution must be exercised to avoid disrupting the cultures and detaching the cells.

Protocol 3. Monitoring cell synchrony

Reagents

- [^{3}H]Thymidine (20 C/mol) (Dupont NEN)
- Appropriate culture medium
- 10% trichloroacetic acid (TCA): prepare 100% TCA; dissolve 500 g of TCA (Sigma) in 227 ml of water. Dilute TCA to 10% and store in an amber glass, covered vessel at 4°C
- Scintillation vials (Fisher)
- 10% sodium dodecyl sulphate (SDS): prepare 10% SDS; dissolve 10 g SDS (Sigma) per 100 g of glass distilled water. Store at room temperature
- Ecolume liquid scintillation counting cocktail (ICN Biomedicals, Inc.)
- Liquid scintillation spectrometer

A. *Determination of DNA synthesis rate in suspension cultures*

1. Add [^{3}H]thymidine to 2 ml of cells to a final concentration of 5 μCi/ml and incubate the cells at 37°C for 30 min with gentle agitation.
2. Pellet by centrifugation at 600 *g* for 5 min and remove the medium by aspiration.
3. Wash the cell pellet twice in ice-cold serum-free medium followed each time by centrifugation at 4°C at 600 *g* for 5 min.
4. Resuspend the cell pellet in 6 ml of cold 10% TCA and maintain in an ice bath for at least 5 min.
5. Pellet precipitate by centrifugation at 600 *g* for 5 min and remove the medium by aspiration.
6. Repeat steps 4 and 5.
7. Resuspend pellet in 500 μl 10% SDS and transfer to a liquid scintillation vial.
8. Add 16 ml of ecolume liquid scintillation counting cocktail to each vial. Radioactivity levels are quantitated in a liquid scintillation spectrometer.

Protocol 3. *Continued*

B. *DNA synthesis rate in monolayer cultures*

1. Add [^{3}H]thymidine to the culture medium to a final concentration of 5 μCi/ml and incubate cells at 37°C for 30 min.
2. Remove the culture medium by aspiration and rinse the monolayer twice with ice-cold serum-free media.
3. Add 5 ml 10% TCA and keep in an ice bath for a least 5 min.
4. Remove TCA by aspiration.
5. Repeat steps 3 and 4.
6. Add 1 ml 10% SDS and maintain for 2 min at room temperature to solubilize TCA precipitates.
7. Assessment of incorporated radioactivity is carried out as described above (see Step 8 in suspension cultures protocol).

3. Synchronization of continuously dividing cells by mitotic selective detachment

3.1 General considerations

A time-proven protocol for obtaining synchronous cells for investigating cell cycle regulatory parameters is mitotic selective detachment. The basis for this

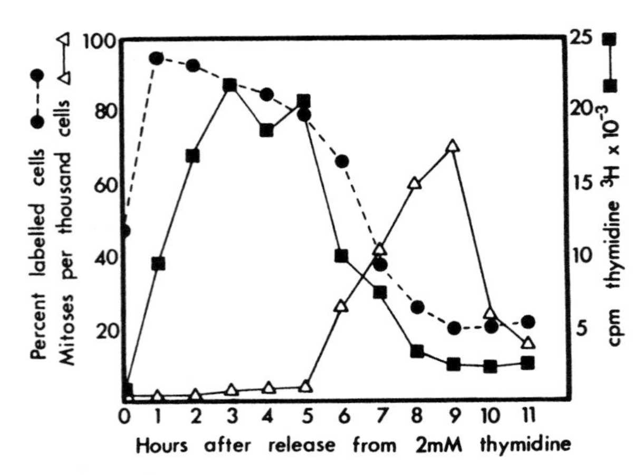

Figure 2. Incorporation of [^{3}H]thymidine into DNA, percentage of cells in DNA synthesis and mitotic index at various times following release of HeLa S cells from a 2 mM thymidine block. Cells were pulsed with 5 μCi/ml of [^{3}H]thymidine for 15 min. The rate of DNA synthesis was determined by the amount of radioactivity incorporated into 5% TCA-precipitable material. The percentage of cells in DNA synthesis and the mitotic index were determined from the autoradiographic preparations.

experimental approach is that transition from a flattened to a spherical shape of cells at the onset of mitosis (prophase) dramatically decreases the area of cell surface attachment. Consequently, mitotic cells can be selectively dislodged from the culture vessel (glass or plastic), harvested by low-speed centrifugation, and replated. Following completion of mitotic division the replated cells spread and synchronously traverse the cell cycle (*Figure 2*). Although the yield of synchronized cells obtained by this experimental approach is low, there are minimal perturbations in biochemical, structural or regulatory parameters of cell cycle regulation or growth control.

The protocol below can be applied routinely for synchronization of suspension cultures of exponentially growing HeLa S3 cells by mitotic selective detachment. Experience dictates that with minimal modifications this approach is readily applicable to many cell types. To optimize the conditions, the entire procedure should be carried out in a 37°C incubator room using all reagents at 37°C.

Protocol 4. Synchronization by detachment of cells in mitosis

Reagents

- Appropriate culture medium

Method

1. Harvest 2.5 litres of exponentially growing HeLa S3 cells at a concentration of 5 × 10^5/ ml in Joklik-modified Eagle's minimal essential medium (MEM) supplied with 7% calf serum by centrifugation at 600 *g* for 5 min. Decant the medium and wash the cell pellet in 200 vol of Eagle's MEM.
2. Resuspend cells in 2000 ml of Eagle's MEM supplemented with 7% calf serum by pipetting repeatedly with a wide bore 25 ml pipet. Monitor by light microscopy for elimination of clumps and continue pipetting until greater than 95% of the cells are a single cell suspension. The switch from Joklik-modified MEM which is calcium-free and contains 10× phosphate to MEM which contains calcium facilitates cell adhesion and cell spreading.
3. Dispense 50 ml aliquots of resuspended cells into 1-litre Blake culture flasks. Place a sterile gauze-wrapped cotton plug in the mouth of the culture flask and place the vessel in a 37°C incubator with a moist 5% CO_2 atmosphere. Alternatively, the flask can be gassed with filter-sterilized 5% CO_2, sealed with a sterile neoprene rubber stopper and placed in a 37°C incubator room. Equally effective for this cell synchronization protocol is placing the cells in 1-litre plastic tissue culture flasks. Cost is the only deterrent to the exclusive use of disposable culture flasks.

Protocol 4. *Continued*

4. At 9 h after cell plating decant the medium gently from the tissue culture vessels and replace it with the Joklik-modified Eagle's MEM supplemented with 7% fetal calf serum. Agitate the vessels 12 times by a gentle rotating motion to selectively dislodge mitotic cells. Pool the media from the vessels and harvest the mitotic cells by centrifugation at 600 *g* for 5 min at 37°C.
5. Decant culture medium carefully and resuspend the cells in 10 ml of Joklik-modified Eagle's MEM supplemented with 7% fetal calf serum. Count the cells in a haemocytometer and evaluate for the percentage of mitotic cells. If more than 95% of the cells are mitotic a satisfactory level of synchrony has been achieved. Dilute the cells to 4×10^5/ml and maintain in suspension culture in a flask with a stirrer. Evaluate cell synchrony by [^{3}H]thymidine labelling and monitor either by determination of TCA-precipitable radioactivity or by autoradiography.

3.2 Modification of mitotic selective detachment protocols

Many normal diploid, transformed, or tumour cell lines can be synchronized by mitotic selective detachment by taking advantage of decreased cell adhesion which occurs during mitosis. Specific modifications in the protocols must be developed for each cell type, but the following general considerations can be effectively applied:

- the synchronization procedure should be initiated with cells as a subconfluent monolayer in exponential growth
- minimal agitation of the culture vessels should be employed for detaching mitotic cells and where possible, modifications of the culture media should be limited
- it is not necessary to maintain the detached mitotic cells in suspension. Mitotic cells obtained by selective detachment from monolayer cultures can be returned to media routinely utilized for monolayer culture, plated and followed throughout the cell cycle as monolayers.

4. Induction of synchronous proliferative activity

The selection and induction protocols thus far described in this chapter have provided methodologies for obtaining synchronized cell populations to define regulatory mechanisms that mediate competency for cell cycle progression. Conceptually, the control mechanisms that can be addressed relate to competency for the onset and execution of DNA replication as well as mitotic division. Parameters of control that contribute to competency for initiation of proliferation, unquestionably associated with responsiveness to growth factors, may involve additional and unique considerations. The initial entry of cells into S-phase following induction of proliferative activity may be somewhat different from those that traverse the G_1/S-phase transition point in continuously dividing cells. Stated within the context of terms frequently employed to describe components of cell cycle and cell growth control, the G_0–G_1 transition or exit from a 'deep' or 'prolonged' G_1 must be further defined at the biochemical, molecular and ultrastructural levels.

To experimentally pursue growth control, defined as the stimulation of quiescent cells to proliferate, there is a series of available options. Quiescence may be achieved by prolonged growth without a change of culture medium or the maintenance of cells in culture media with reduced levels of serum or growth factors (13, 14). Stimulation of proliferation can be achieved by re-feeding serum-deprived cell with fresh medium containing serum by addition of growth factors. An alternative approach that has yielded valuable insights into the requirements for induction of proliferation involves the use of temperature-sensitive mutants of the SV40 virus T-antigen (15, 16 and Chapter 12).

Specific protocols will not be presented in this chapter since they are comprehensively described in the above referenced articles. Synchrony can be monitored by the method previously detailed in this chapter.

References

1. Bootsma, D., Budke, L., and Vos, O. (1963). *Exp. Cell Res.*, **33**, 301.
2. Terasima, T. and Tolmach, I. J. (1963). *Exp. Cell Res.*, **30**, 344
3. Stein, G. S. and Borun, T. W. (1972). *J. Cell Biol.*, **52**, 292.
4. Holthuis, J., Owen, T. A., van Wijnen, A. J., Wright, K. L., Ramsey-Ewing, A., Kennedy, M. B., Carter, R., Cosenza, S. C., Soprano, K. J., Lian, J. B., Stein, J. L., and Stein, G. S. (1990). *Science*, **247**, 1454.
5. Studzinski, G. P. and Lambert, W. C. (1968). *In Vitro*, **4**, 139.
6. Studzinski, G. P. and Lambert, W. C. (1969). *J. Cell. Physiol.*, **73**, 109.
7. Lambert, W. C. and Studzinski, G. P. (1969). *J. Cell. Physiol.*, **73**, 261.
8. Churchill, J. R. and Studzinski, G. P. (1970). *J. Cell. Physiol.*, **75**, 297.
9. Plumb, M., Stein, J. L., and Stein, G. S. (1983). *Nucleic Acids Res.*, **11**, 2391.
10. Baserga, R. and Malamud, D. (1969). *Autoradiography*, Hoeber, New York.
11. Pockwinse, S. M., Wilming, L. G., Conlon, D. M., Stein, G. S., and Lian, J. B. (1992). *J. Cell Biochem.*, **49**, 310.
12. Pedrali-Noy, G., Spadari, S., Miller-Faures, A., Miller, A. O. A., Kruppa, J., and Koch, G. (1980). *Nucleic Acids Res.*, **8**, 377.
13. Wiebel, F. and Baserga, R. (1969). *J. Cell. Physiol.*, **74**, 191.
14. Rossini, M., Lin, J.-D., and Baserga, R. (1976). *J. Cell. Physiol.*, **88**, 1.
15. Ashihara, T., Chang, S. D., and Baserga, R. (1978). *J. Cell. Physiol.*, **96**, 15.
16. Soprano, K. J., Dev, V. G., Croce, C. M., and Baserga, R. (1979). *Proc. Natl. Acad. Sci. USA*, **76**, 3885.

11

Growth and activation of human leukaemic cells *in vitro* and their growth in the SCID mouse model

SOPHIE VISONNEAU, ALESSANDRA CESANO, and DANIELA SANTOLI

1. Introduction

Despite their aggressive growth in the patient's body, leukaemic cells usually fail to proliferate autonomously in tissue culture conditions (1). The dependence of human leukaemic cells on specific haematopoietic growth factors for *in vitro* growth was first demonstrated in well- or partially-differentiated acute T lymphoblastic leukaemias (T-ALL) which could be maintained in conditioned media containing interleukin 2 (IL-2) (2). Many of these cell lines have subsequently become growth factor-independent due to autocrine production of IL-2. The growth factors required by acute myelogenous leukaemias (AML) for *in vitro* growth have been defined more recently. Initially, it was observed that a small fraction of AML cells could form colonies in semi-solid medium supplemented with conditioned media from normal myeloid cells (3–5), or from phytohaemagglutinin-stimulated lymphocytes (6). It was later found that IL-3 and granulocyte–macrophage colony-stimulating factor (GM-CSF) represented the active factors in these preparations (7, 8).

In the past decade, we have investigated the growth factor requirements of more than 100 cases of human leukaemias, using recombinant cytokine preparations provided by Dr Steven C. Clark (Genetics Institute, Cambridge, MA). Sections 2 and 3 of this chapter summarize the most salient findings of these studies and detail the strategy used successfully to expand leukaemic cells in suspension cultures and semi-solid medium, respectively. During the process of characterization of the T-cell lines we established from children with T-ALL, we found that three of them were endowed with tumoricidal activity and ability to produce immunomodulatory lymphokines. A great deal of work has consequently been done in this laboratory to study these IL-2-dependent killer clones, as compared to cytokine-activated peripheral blood lymphocytes (PBL) from healthy donors. Section 5 is dedicated to a thorough description of the methods used to induce and measure cytokine production and cytotoxicity in T-ALL cells and PBL.

Section 4 describes our procedures for engrafting human leukaemic cells in severe combined immunodeficient (SCID) mice. Using this approach, we have found that (a) human leukaemia cells disseminate in SCID mouse tissues inducing a clinical picture reminiscent of the patient disease; (b) the cells recovered from the infiltrated organs are karyotypically, genotypically, and phenotypically identical to the original cells injected; and (c) in many instances, primary leukaemic cells that cannot be maintained in tissue culture conditions are able to grow in the SCID mouse and can be serially propagated in these animals. The latter finding indicates that the SCID mouse model provides a more suitable environment for the growth of human leukaemic cells than *in vitro* conditions. The possibility of growing primary patient cells in SCID mice not only allows to generate large quantities of leukaemic cells for various types of studies, but has immense prognostic, diagnostic, and therapeutic potential as well.

2. Procedures for growing human leukaemic cells in suspension culture

The recombinant cytokines mostly used in our studies to culture leukaemic cells are: IL-1α, IL-2, IL-3, IL-4, IL-5, IL-6, IL-7, IL-11, stem cell factor (SCF) (c-*kit*), GM-CSF, and G-CSF. Based on its ability to promote the growth of very immature haematopoietic progenitors, IL-3 was found to be the major inducer of both AML and ALL cell proliferation. In general, IL-2 appeared to be the optimal growth stimulatory cytokine for T-ALL cells, with the exception of highly immature T-ALLs that responded only to IL-3 or GM-CSF as the initial stimuli. In some cases, GM-CSF and IL-5 supported the growth of AML cells independently from IL-3. Moreover, IL-6, IL-7, IL-11, or SCF were often able to stimulate the initial growth of both lymphoid and myeloid leukaemias, especially when combined with IL-3. Although most AML and ALL samples underwent terminal differentiation into mature elements and stopped dividing shortly after culture in optimal cytokine combinations, we did have success in immortalizing leukaemic samples from several paediatric cases of leukaemia (9–11). Some of the established pre-B ALL or T-ALL cell lines have become growth-factor independent with time, whereas other cell lines derived from AML or T-ALL cases still require IL-3, GM-CSF, or IL-2 for viability and continuous growth. The phenotypical, functional, genotypical, and karyotypical properties of these cell lines have been reported in detail (9–11). This section provides a general strategy for growing leukaemic cells in long-term cultures, using recombinant haematopoietic cytokines.

Protocol 1. Fractionation of leukaemic bone marrow (BM) and peripheral blood (PB) samples

PB (2–10 ml) and BM (0.5–1.0 ml) samples are provided by the collaborating physicians in heparinized glass tubes or plastic syringes; heparin is used to prevent clotting. It is important to use sterile reagents and supplies throughout the fractionation procedure to avoid contamination.

1. Dilute the sample in Iscove's modified Dulbecco's medium (IMDM; Gibco) supplemented with 10% heat-inactivated fetal bovine serum (FBS; Sigma), and antibiotics (penicillin and streptomycin) (Gibco).
2. Layer the leukaemic sample (slowly and carefully) on a Ficoll–Paque (F-P) solution (Pharmacia LKB) at the ratio of 2:1, in a plastic centrifuge tube (Falcon). Because F-P is stored at 4°C, the required amount of this reagent must be warmed up in a water bath at 37°C at the time of use.
3. Centrifuge the tube at room temperature (18–20°C) at 800 *g* for 30 min. The resulting gradient will have an interface containing mononuclear low density cells (lymphocytes, monocytes, stem cells) and a pellet containing erythrocytes, granulocytes, and most of the metamyelocytes. Pick up the interface with a plugged Pasteur pipette, transfer it into a plastic tube, and wash away the F-H solution by centrifugating twice in phosphate-buffered saline (PBS) (450 *g* for 10 min at 4°C).
4. Resuspend the pellet in a small volume (usually 2.0 ml) of medium and count the cells by diluting 1:10 or 1:20 in 1% erythrosin B (Sigma) to monitor their viability. Depending on the type of leukaemia and disease stage, different cell counts will be obtained. A 1 ml diagnostic sample usually provides $2–10 \times 10^7$ cells, mostly representing leukaemic blasts. The few normal lymphoid and myeloid elements present in these samples will survive for a short time in culture conditions.
5. Cryopreserve most of the leukaemic cells in liquid nitrogen at $1–2 \times 10^7$ cells/ml/vial, using a freezing solution made up of 40% medium, 50% FBS, and 10% dimethyl sulphoxide (DMSO; Sigma).
6. Plate the rest of the cells directly in 24-well culture plates (Falcon) (2×10^6/2 ml/well) in the presence of recombinant growth factors (see below); a control well will receive medium alone to monitor the autonomous growth of the leukaemic sample. Growth factor-responsive, rapidly growing cells can soon be transferred into a 6-well plate (5 ml/well) (Falcon), and later into T25 (10 ml) and T75 (25 ml) flasks (Falcon) for further expansion and cryopreservation.

By first performing short-term proliferation assays on freshly separated leukaemic cells in the presence of single and combined recombinant growth factor preparations (*Protocol 2*), it is possible to identify the optimal cytokine combination(s) and dosage(s) to be used for long-term culture of each leukaemic sample, thus increasing the chance of successfully expanding the cells *in vitro*. The cryopreserved leukaemic cells can then be thawed and plated in wells or flasks, according to *Protocol 1*. It is imperative to culture leukaemic cells at a high concentration (at least 1×10^6/ml), regardless of the vessel in which they are incubated. The need by these cells to stay in close contact may reflect an autocrine production of growth stimulatory cytokines. If well-growing and healthy leukaemic cells are over-diluted when passaged, they will aggregate, produce toxic debris, stop growing, and finally die. Leukaemic cell cultures are commonly passaged every 3–4 days by adding the right amount of medium and growth factor. We have found that human haematopoietic cells (both normal and leukaemic) grow best at 37°C in a humidified incubator containing 8–10% CO_2: several of our leukaemic cell lines, especially T-ALL, stop dividing when the CO_2 level is kept at 5% but often revert to a brisk growth when the CO_2 is increased to 10%.

Protocol 2. Short-term proliferation assay

1. Suspend the cells at 1×10^6 cells/ml in complete medium and dispense them at 100 µl/well into a flat-bottomed 96-well microplate (Falcon).
2. Dilute the recombinant haematopoietic growth factors in complete medium to give 3–5 different concentrations and add them (100 µl/well) to the cells. Make sure to have triplicate wells per growth factor concentration. Control wells will receive 100 µl medium to measure the proliferative ability of the leukaemic cells in the absence of exogenous cytokines.
3. Incubate the plates for 3 days at 37°C in a humidified 10% CO_2 incubator.
4. Add 1.0 µCi/well of [^{3}H]methyl-thymidine (^{3}H-TdR) (6.7 Ci/mM; Amersham or ICN Biomedicals).
5. At 6–18 h later, harvest the cells using an automated multi-well harvester that aspirates and lyses the cells and transfers DNA on to filter paper (Packard), while allowing unincorporated ^{3}H-TdR to be washed out.
6. Let the filters dry, and count the radioactivity in a scintillation counter until standard deviation is $<2\%$. Calculate mean c.p.m. for each experimental condition. There should be less than 10% variation in triplicate cultures.

3. Clonal growth of leukaemic cells in semi-solid medium

Clonal culture of haematopoietic cells is usually performed to: (a) identify and enumerate haematopoietic precursors; (b) determine the frequency and lineage of progenitor cells in a given leukaemic sample; and (c) identify soluble or cellular factors that affect the proliferation/differentiation of normal and neoplastic haematopoietic cells. Although the methods for preparing and scoring clonal cultures are similar for normal and leukaemic cells, the specific protocols may differ depending on the objective of the clonogenic assay. The general procedure is to mix tissue culture medium with agar or methylcellulose, add the leukaemic cells, and dispense the resulting cell suspension into Petri dishes. In most cases, formation of colonies by human leukaemic samples is not autonomous and requires addition of appropriate growth factors (12).

3.1 Choice of the semi-solid medium

A wide variety of tissue culture media have been successfully used to support leukaemic colony formation *in vitro*. We routinely use IMDM supplemented with 10% FBS, unless the experimental protocol requires cloning in serum-free conditions. Clonogenic cultures are rendered semi-solid by inclusion in the culture medium of either agar or methylcellulose. The cloning efficiency of all major subtypes of haematopoietic colonies is identical in either condition, with the exception of erythroid colony growth (CFU-E): erythroid cells appear to proliferate better in methylcellulose cultures and overall frequencies of CFU-E in agar are only 1/3–1/4 of those in parallel methylcellulose cultures. On the other hand, agar is preferable in most instances, based on the following.

(i) The preparation of agar cultures is less time consuming than that of methylcellulose cultures.
(ii) At their final concentrations, agar is a true gel, whereas methylcellulose is a viscous liquid. As a consequence, clones grown in agar remain more tightly packed and, more importantly, remain fixed in location in a three-dimensional matrix. By contrast, clones in methylcellulose sediment slowly during the incubation period and often become superimposed one over the other near the bottom of the dish. This phenomenon can completely prevent scoring of small size colonies.
(iii) For the same reason, removal of clones from agar can usually be accomplished without disturbing the adjacent clones, whereas removal of single clones from methylcellulose can be accomplished only if the clones are widely separated from each other.
(iv) Fixation of agar cultures is more simple (see Section 3.2.1) than fixation of methylcellulose cultures (13, 14).

3.2 Preparation of the agar stock

Protocol 3. Preparation of agar stock

1. Transfer 3 g of agar (Bacto-agar from Difco, or Agar noble from Sigma) into a Pyrex glass bottle containing 100 ml of dionized distilled water.

Protocol 3. *Continued*

2. Close the bottle with a loose cap and boil in water for 20–30 min to dissolve and sterilize the agar.
3. Shake the bottle every 5 min until the agar becomes fluid, dispense it into plastic tubes, glass bottles, or vials.
4. Store these aliquots of 3% agar for a maximum of 3–4 months at room temperature. Each aliquot can be boiled several times.

Protocol 4. Plating of leukaemic cells in agar

As for the growth of leukaemic cells in suspension culture, sterile conditions have to be used throughout the procedures of plating and growing the cells in semi-solid medium.

1. Label 35 × 10 mm plastic Petri dishes with a marker (prepare triplicate dishes per experimental condition).
2. Prepare the mixture containing cells, medium, and the growth factor preparation(s) in plastic 15 ml tubes (Falcon).
3. Boil (in water for 20 min) an aliquot of prepared 3% agar stock solution (*Protocol 3*) and set it at 37°C for a few min before using it. Then add the proper amount of agar to the cell mixture to give a final concentration of 0.3% agar. Transfer 1 ml of the agar–cell mixture to each Petri dish.
4. Allow the cultures to gel on a bench that is cool and vibration-free.
5. Transfer the Petri dishes into a bigger tray and then gently into a CO_2 incubator.
6. Score the number and size of colonies 10–14 days later.

Note: generally, when the clonogenic frequency is expected to be low (such as in the case of primary leukaemic samples) as many as $2–4 \times 10^5$/ml should be plated. In instances in which a high cloning efficiency is expected (typical of several leukaemic cell lines), it is important to culture the cells at very low density (10^4 or even 10^3 cells/ml) to avoid crowding and overlapping of the colonies.

3.2.1 Fixation of colonies in agar

Colonies in agar cultures can be fixed to fully preserve their shape and orientation. Conveniently, once fixed, the agar colonies can be scored even 1 month later.

(i) Use 2.5% glutaraldehyde in PBS as fixative (prepare the stock and keep it at room temperature away from light); add 4–5 drops of this solution/Petri dish.

(ii) After 1 h incubation at room temperature, add 0.3 ml of tap water/dish.

(iii) Cover the dishes and transfer them on to a large tray containing a small Petri dish filled with water (to maintain moisture). Cover the tray and store it at 4°C until ready to read the results.

3.3 Preparation of the methylcellulose stock

The preparation of methylcellulose cultures is slower and more tedious than that of agar cultures. However, it is easier to obtain good cytological preparations from clones removed from methylcellulose since the medium readily becomes less viscous on addition of extra fluids, and the cells can simply be dispersed on a microscope slide using an air jet. It is difficult to smear agar colonies without disrupting many colony cells. The use of methylcellulose is recommended in special situations involving CFU-E analysis or the recloning of mixed colonies where a detailed morphological study needs to be performed (14). We found that in addition to erythroid cells, T cells also clone better in methylcellulose than in agar.

Protocol 5. Preparation of methylcellulose stock

1. Weigh 2.5 ml of methylcellulose (Dow) and put it with a magnetic stirrer in a Pyrex bottle containing 100 ml of IMDM without serum.
2. Stir, close the bottle with a loose cap and autoclave it.
3. Transfer the bottle on to a magnetic plate (don't forget to tighten the cap) and let the solution stir till it appears homogeneously viscous (48–72 h).
4. Store at 4°C.

3.3.1 Plating the cells in methylcellulose

Mix the cells, serum, and growth factors as described in *Protocol 4*. Add the 2.5% methylcellulose stock solution to have a 0.8–1% final concentration. The following suggestions are given for better results.

(i) Whatever the final volume, prepare 0.5–1.0 ml in excess of what is needed, because the cell suspension, after adding the methylcellulose, is very sticky.

(ii) Because of its high viscosity, the methylcellulose stock solution can be picked up only by using a tuberculin syringe without needle.

(iii) Always add the methylcellulose to the plastic tube as the last 'ingredient' by dropping it in the centre of the tube. Avoid forming drops on the walls of the tube as they are very difficult to dissolve. Change syringe every two tubes.

(iv) Cap the tubes tightly and vortex the samples until the mixture is homogeneous.

(v) Use a tuberculin syringe with a 18G1/2 needle to transfer the solution from the tube to the Petri dishes.

(vi) Transfer each individual dish in the incubator as soon as it is filled.

3.4 Scoring of colonies in semi-solid medium

Colonies are scored under an inverted microscope or a dissecting microscope with a light source in the base plate. Counting colonies in a 35 mm Petri dish is best performed by placing the culture dish on the bottom of an inverted 50 mm culture dish on which a convenient grid pattern has been ruled (approximately 5 mm between the lines), using a black marker: the two dishes are then moved as a single unit to count colonies in vertical stripes. A large portion of colonies is usually found at the edges of the cultures, partly because of the centrifugal motion used to mix the cultures, and partly because of a meniscus effect that causes the cultures to be thicker at the edges. Care needs to be taken not to miss these colonies at the edge of the culture dish. Colonies are conventionally defined as containing 50 or more cells. Clones containing less than 50 cells are designated 'clusters' or 'aggregates'.

3.5 Staining of colony-forming cells

Several good staining procedures for mouse haematopoietic colonies are available, such as orcein staining, acetylcholinesterase stain (for megakaryocytes), the use of benzidine-based stains (to identify erythroid cells), and Astra Blue (for mast cells). For human haematopoietic colonies, the May Grünwald–Giemsa stain on cytocentrifuged cell preparations fixed with methanol is an ideal all-purpose stain. Luxol Fast Blue followed by haematoxylin is a very useful stain combination either for picked-off colonies or whole culture preparations to identify erythroid, granulocyte–macrophage, eosinophil, and megakaryocytic cells. Luxol Fast Blue is a specific stain for eosinophils which contain large, intensely green, cytoplasmic granules. Counterstaining with haematoxylin allows the identification of other cell types by nuclear morphology. Other combination stains are based on the histochemical demonstration of different esterase activity in human granulocytes and monocytes using two different substrates. The details of good staining procedures for haematopoietic colonies are summarized in ref. 14 and can be found in more recent literature.

4. Use of the SCID mouse model for growing human leukaemias

4.1 The SCID mouse

Mutant SCID mice are homozygous for an autosomal recessive mutation on chromosome 16 and have a defective VDJ recombinase system, necessary for the rearrangement of antigen-receptor genes on both T and B cells (15). As a consequence, these mice lack mature and functional T and B cells. Despite the impairment in lymphocyte differentiation, the SCID mouse haematopoietic microenvironment, including the thymic and BM stroma, is intact, as it contains normally functioning antigen-presenting cells and is able to promote the differentiation of normal stem cells into functionally competent T and B lymphocytes. These mice also have a normal natural killer (NK) cell function. Since its discovery in 1983, the SCID mouse has been used increasingly as a model for the murine haematopoietic and immune systems, and as recipient of human grafts, including leukaemias (16–19). This section will review step by step the procedures we have used to engraft human AML and ALL samples in SCID mice, and will provide specific tips for a successful approach.

4.1.1 Determination of plasma levels of IgM in SCID mice

SCID mice are commercially available from either Taconic or Charles River. They have to be housed in a pathogen-free environment provided by autoclaved, microinsulated cages. Their litter, food, and water must also be autoclaved. Mice have to be manipulated exclusively under laminar-flow hoods. If all these precautions are taken, antibiotic prophylaxis is not needed.

Six- to 8-week-old mice (of either sex) are used for engrafting human leukaemias because, as a function of age and exposure to environmental stimuli, their lymphoid progenitor cells often pass through the differentiation barrier to produce mature T- and B-cell progeny ('leaky' SCID mice). A good indicator of leakiness is the plasma level of IgM (the first isotype of Ig that appears during a humoral immune response). The enzyme-linked immunosorbent assay (ELISA) (see *Protocol 6*) is an accurate, quick, and inexpensive method to measure IgM levels in the SCID mouse plasma. Serum IgM-positive mice should not be used as they are bad recipients of human leukaemias (absence or delay in leukaemic cell engraftment has occurred in our studies with these mice).

4.1.2 Obtaining and testing plasma samples

We draw blood by orbital puncture using heparinized blood collecting microglass tubes (full tube volume is 250 μl) (Fisher Scientific); each tube is then placed inside a conical 15 ml plastic tube (Becton Dickinson) and centrifuged at 800 *g* for 20 min. After centrifugation, a clear separation between erythrocytes (pellet), white blood cells (thin white ring), and plasma (top fraction) is obtained. At this point, the plasma can be collected by cutting the tube with a blade, at 1–2 mm above the white ring, and by emptying the plasma-containing segment into an Eppendorf tube.

Protocol 6. Determination of plasma levels of IgM by ELISA

Equipment and reagents

- 96-well/U-bottomed vinyl plate (Costar)
- 0.1 M carbonate/bicarbonate buffer, pH 9.6
- Wash buffer: PBS + 0.05% Tween 20
- Normal mouse IgM (Sigma) as standard serum
- Monoclonal antibody (mAb) anti-mouse IgM (Cappel)
- mAb anti-mouse IgM-peroxidase conjugated (Cappel)
- Colorimetric substrate (Kirkegaard and Perry Lab.) consisting of a solution A (ABTS peroxidase substrate) and a solution B (peroxidase)

Method

1. Dilute the mAb anti-mouse IgM in 0.1 M buffer to a final concentration of 1 μg/ml.
2. Put the diluted mAb into the plate (100 μl/well) and incubate overnight at 4°C.
3. Wash the plate five times: fill each well with wash buffer and remove the liquid by inverting the plate and blotting it against a clean paper towel.
4. Dilute the experimental plasmas 1:100 in the wash buffer. A suggested dilution series for the standard IgM is 3, 1, 0.3, 0.1, 0.03, 0.01, 0.003, and 0.001 μg/ml.
5. Add to the plates the standard and diluted plasma samples (100 μl/well, in duplicate wells) and incubate 4 h at room temperature.
6. Repeat Step 3.
7. Add 100 μl/well of mAb-peroxidase conjugated (diluted 1:3000 in wash buffer) and incubate in the dark for 1 h.
8. Repeat Step 3.
9. Mix the colorimetric substrate solutions A and B in equal amounts in a glass container and add 100 μl of this substrate into each well.
10. Let the colour develop for 30 min and then read the plate with an ELISA reader at a wavelength of 405 nm. Use, as cut off for positivity, an absorbance value that is three times the value of the buffer alone.

4.2 Immunosuppressive treatments of SCID mice

In our experience, the incidence of successful engraftment of human leukaemias in SCID mice is increased by further immunosuppressing the animals with antiblastic drugs, such as Etoposide (VePesid®, Bristol Laboratories). We inject this drug intraperitoneally (i.p.; 100 mg/kg of body weight in 0.5 ml PBS), 3–4 days before transferring the leukaemic cells. This is the length of time required to impair the natural immunity of the SCID mice, which is mediated by NK cells and macrophages, and to 'clear' the mouse body of active drug metabolites; if still present at the time of engraftment, these metabolites can impair the viability of the leukaemic cells. Another way to further immunosuppress the mice is γ-irradiation provided by a ^{137}Cs source. Because the enzymatic defect responsible for the lack of mature lymphocytes in SCID mice also affects their capability to repair DNA damage (20), 200 rads is the maximal dose that should be given to these animals. When irradiation is used as the immunosuppressive method, the leukaemic cells should be injected immediately thereafter. Thus, irradiation offers the advantage over drug treatment in that patient samples can be injected immediately after their separation, when cell viability is optimal and the percentage of leukaemic blasts is very high.

4.3 Transfer of leukaemic cells i.p.

The number of cells transferable i.p. can be as high as 5×10^7 (we usually inject a maximum of 2×10^7 cells). The cells are first washed, resuspended in PBS at the final volume of 500 μl/mouse, and injected using a 1 ml tuberculin syringe with a 27G1/2 needle. The use of a larger needle is contraindicated because it produces a bigger hole in the abdomen, facilitating the leakage of the injected liquid.

4.4 Intravenous (i.v.) injection of leukaemic cells

To dilate and better visualize the dorsal tail vein, we expose the tail to the heat of an infrared lamp or immerse it in warm water for a few minutes. Cells are injected at approximately the mid-point of the length of the tail, using a 1 ml tuberculin syringe with a 27G1/2 needle. The maximum number of cells transferable i.v. is inversely proportional to the size of the leukaemic cells. In fact, a complication that can occur during i.v. injection is acute right-side heart failure due to a sudden increase in the lung resistance following pulmonary embolism; this complication increases proportionally with the number and the size of the cells injected, and is almost always fatal. To avoid this problem, we recommend the injection of a number of cells $\leqslant 1 \times 10^7$ in a volume of PBS $\leqslant$ 100–200 μl.

4.5 Success of human leukaemic cell engraftment

We have had close to 100% success in engrafting T-ALL samples in SCID mice and 50–60% success in engrafting AML and pre-B-ALL samples. A likely explanation for the difficulty in growing myelogenous leukaemias in SCID mice is the tendency of such leukaemias to terminally differentiate into monocytic and granulocytic elements (Section 2). Recently, a new chimeric SCID mouse model has been described (21, 22), which allows 100% of AML cell engraftment. In this model, human fetal bone fragments are implanted subcutaneously and the leukaemic cells are injected directly into the bone marrow grafts, thus encountering an environment similar to that in the patient's marrow. This procedure is sophisticated and technically demanding, but provides a useful tool for expanding leukaemic cells and for detection of residual leukaemia in remission marrows.

4.6 Appearance of symptoms

Each type of leukaemia induces distinct patterns of infiltration in the SCID mouse lymphoid and non-lymphoid organs, which is often highly reminiscent of the clinical findings in the patients. The delay between the injection of the leukaemic cells and the appearance of symptoms is very different from one leukaemia to another, varying from a few weeks to several months. However, in most cases, the mice die shortly after appearance of symptoms. We strongly recommend a daily monitoring of the condition of the mice because histopathology and cell recovery performed on mice dead for longer than 4 or 5 h usually give poor results. Signs indicative of disease are:

- ruffled fur which is often the first sign of distress
- progressive enlargement of the abdomen, due to growing abdominal tumour masses, splenomegaly, hepatomegaly, or ascites formation
- respiratory distress (when present in a mouse without enlarged abdomen, it frequently indicates the presence of pleural effusions and/or mediastinal lymphoadenopathy)
- hunched posture due to respiratory distress
- neurological symptoms (gait disturbance, hind limb weakness and paralysis are signs of spinal cord involvement and appear with some types of leukaemias only when the cells are injected i.v.)
- lethargy (the mouse poses little if any resistance to the investigator's manipulations; in some cases, lethargy indicates central nervous system involvement)

4.7 Autopsy and histopathology

We sacrifice the mice when they show advanced signs of illness or heavy tumour burden. Mice can be euthanized using different methods: most commonly, drugs (overdose of anaesthetic), carbon dioxide, or cervical dislocation. The last should be avoided when collection of brain tissue and spinal cord are required because these organs are damaged during this procedure. Autopsy is performed on mice placed on their backs on clean, dry, absorbent paper. Wetting the fur with 70% ethanol reduces the chance of contamination. After a midline incision with a sterile scalpel and retraction of the skin above chest and abdomen, proceed as follows:

(i) To recover ascites (most of the time ascites contain a high number of leukaemic cells), extra attention has to be paid not to cut the peritoneum before all the liquid is aspirated. Use a tuberculin syringe (with a 23G needle) side-positioned on the lower abdomen.

(ii) To recover pleural effusions, insert a tuberculin syringe (with a 23G needle) into the mouse chest between the ninth and tenth rib on both sides and aspirate the liquid.

(iii) To collect PB, insert a tuberculin heparinized syringe perpendicularly in the mouse chest (2–4 mm) between the seventh and eighth rib on the left side: once the needle is correctly positioned in the heart, 0.5–1 ml of PB can be easily drawn.

(iv) BM samples are collected from the mouse femoral bones. First remove the surrounding muscles with surgical scissors. Cut the bone perpendicularly at the level of the pelvic articulation; insert a 27G1/2 needle of a syringe containing 200–300 μl of medium, and aspirate the BM with strength. Because the BM cells will remain inside the needle, their recovery is achieved by emptying the content of the syringe into a culture tube. Using this technique, $2–3 \times 10^6$ cells per femur can be collected.

(v) For a 'clean' collection of the abdominal organs, the following order is suggested: spleen (which is usually enlarged), tumour masses (often originating from the mesenteric lymph nodes), liver (sometimes there is involvement of the lymph nodes around the hilus hepaticus with gall bladder enlargement), kidneys (frequently surrounded by lymph nodal retroperitoneal tumour masses), ovaries or testes.

(vi) The removal of the lung–heart–thymus complex is facilitated by a longitudinal cut with surgical scissors beginning at the xiphoid and extending to the neck.

(vii) The brain can be removed by cutting in a circle, with surgical curved scissors, the top part of the skull (from one ear to the other). Once the brain is exposed, its removal is facilitated by the use of curved forceps inserted at the base of the brain.

Immediately after removal, the organs should be placed in 5–10 ml of tissue culture medium in sterile Petri dishes. For histopathological analysis, the tissues should be fixed for at least 24 h in 10% formalin, and paraffin embedded; 4 μm sections are then cut and stained with haematoxylin and eosin (Sigma).

4.8 Recovery of human leukaemic cells from mouse tissues

(i) *Pathologic fluids and BM*: after collection in plastic tubes, centrifuge these samples at 450 *g* for 10 min, discard the supernatant, and resuspend the pellets in culture medium.

(ii) *PB*: dilute the blood 1:1 with PBS, centrifuge it on a F-P gradient at 450 *g* for 30 min, and collect the ring at the interface. Wash the cells twice in PBS and resuspend them in culture medium.

(iii) *Spleen, liver, thymus, lymph nodes, etc.*: mince the organs through sterile metal grids (placed on a culture dish) while adding medium with a syringe piston. Centrifuge on a F-P gradient to remove dead cells and erythrocytes; collect the interface and proceed as for the PB.

Upon recovery from mouse tissues, it is recommended that the karyotypic and phenotypic properties of the leukaemic cells are analysed to determine whether the original chromosomal translocations and lineage-specific surface markers were retained or whether a leukaemic subclone was preferentially expanded. In the case of T-ALL cells we also perform Southern blot analysis to confirm the presence of the original T-cell receptor rearrangements. In addition to culturing the cells in the presence and absence of recombinant haematopoietic growth factors, we transfer the cells into a new group of IgM-negative SCID mice in order to perpetuate their growth *in vivo* in case the *in vitro* conditions are not adequate.

5. Human lymphocyte activation in culture

In this section, we will describe the protocols most commonly used in our laboratory to measure lymphocyte activation as applied to both conventional PBL, NK cells, and lymphokine-activated killer (LAK) cells.

5.1 Cell-mediated cytotoxicity

The lytic event whereby human lymphocytes kill target cells is composed of three distinct phases: (i) the recognition phase, in which effectors and targets come into close contact (binding step); (ii) the lytic phase (Ca^{2+}-dependent degranulation and release of lytic proteins), in which the effector delivers the lethal hit; and (iii) the final phase, in which the target cell undergoes necrotic or apoptotic changes and dies. Different types of receptors are involved in the effector–target (E/T) binding event, depending on whether specific CTL or major histocompatibility complex (MHC) non-restricted effectors (NK and LAK cells) are involved. Necrotic changes in the target are typically characterized by blebbing and cell swelling without nuclear involvement resulting in osmotic death. Morphological changes such as nuclear degradation with dense aggregates of chromatin and invagination of the nuclear membrane (with no changes in the cytoplasm or cell membrane) are instead indicative of cell death by apoptosis (also named 'programmed cell death' or 'induced suicide', see Chapter 7). The two mechanisms of lysis, pore formation produced by secreted perforin (PFP) and contact-induced triggering of internal disintegration (apoptosis), are not mutually exclusive, and may be used by the same effector to lyse different types of tumour targets.

5.2 Expression of intracytoplasmic azurophilic granules and lytic proteins

Most types of cytotoxic lymphocytes, including CTL, NK, and LAK cells, express cytoplasmic azurophilic granules containing lytic proteins, such as PFP, serine esterases (SE; also called granzymes) (23–26), and a protein involved in apoptosis named TIA-1 (27). These proteins play an essential role in the post-binding phase of lytic interaction leading to disintegration of the target. Cytotoxic granules can be constitutive or induced by cytokines, such as IL-2 and IL-12 (28). The presence of granules in the cytoplasm of a lymphocyte preparation can be visualized by morphological analysis. With this analysis it is possible to evaluate the ability of a given stimulus to induce granule formation, and correlate the expression of the intracytoplasmic 'lytic machinery' with the extent of cytotoxic activity. Once the cells are concentrated on slides by cyto-centrifugation, they are stained with May Grünwald–Giemsa, as follows:

(i) Cover the slides with undiluted May Grünwald solution (EM Diagnostic Systems) for exactly 6 min.
(ii) Wash extensively with distilled water and add Giemsa (Accra Lab) (freshly diluted 1:10 in water) for 7 min.
(iii) Wash again with water.
(iv) Let the slides dry vertically on a paper towel and examine them under a microscope.

To quantitate the levels of expression of cytotoxic components stored in the cytoplasmic granules, immunochemical techniques are available (28). For immunoperoxidase or alkaline-phosphatase staining of PFP and TIA-1, we use commercially available kits (Vectastain ABC Kits, Vector Laboratories), which contain the necessary reagents. The cells are fixed in cold acetone for 10 min immediately after cyto-centrifugation, and stained with an anti-TIA-1 or anti-PFP antibody. This procedure is described in detail in ref. 28. The method for detection of SE1 is described in *Protocol 12*.

5.3 E/T conjugate formation

By counting the number of E/T conjugates one can determine whether an increase in the lytic activity of a given cytotoxic population (e.g., upon activation with cytokines or other stimuli) can be attributed to an increased capability of the effectors to bind to the targets. We quantitate E/T conjugates using the method of Callewaert *et al.* (29).

Protocol 7. E/T conjugate formation

1. Wash the effector cells (1×10^6) three times in PBS at 4°C, resuspend them in 1 ml PBS containing 1% bovine serum albumin (pH 7.4), and incubate with Calcein-AM (final concentration 1 μg/ml) (Molecular Probes) for 1 h at room temperature in the dark.
2. Wash the target cells (1×10^6) three times in PBS at 4°C, and label them with hydroethidine (Molecular Probes) in PBS–bovine serum albumin (final concentration 40 μg/ml) for 1 h in the dark.
3. Wash effectors and targets (separately) three times with ice-cold PBS and resuspend them in PBS containing 5 mM MgCl and 1 mM EGTA at 4°C.
4. Mix effectors and targets (1:1 ratio) and centrifuge at 100 *g* for 3 min. Transfer the tubes to a 37°C incubator for 10 min, and gently resuspend the conjugates in ice-cold medium.
5. Analyse conjugated and unconjugated cells by dual-colour flow cytometry (Chapter 2) using an Epics Elite machine (Coulter Corporation) with an excitation wavelength of 488 nWatt, 15 mV, and green (525 BP) and red (575 BP) emission filters (gates have to be set in each experiment to optimize the separation of the two fluorescent signals). A total of 10 000 events have to be counted for accuracy. The percentage of conjugated cells is calculated by dividing the number of dual-labelled particles by the total number of effectors and multiplying by 100.

5.4 Direct lysis of target cells

The basic procedure to measure lymphocyte-mediated cytotoxicity is the ^{51}Cr ($Na_2{}^{51}CrO_4$) release assay (10, 11, 30).

Protocol 8. ^{51}Cr-release assay

The E/T ratios may vary considerably in each assay, depending on the potency of the effector cell and the susceptibility to lysis of the target cell. The E/T ratios exemplified below are 100:1, 50:1, 25:1, and 12.5:1. Being a short assay, it is not necessary to use sterile conditions.

1. Spin the required number of target cells at 450 *g* for 5 min in a 15 ml conical tube, aspirate the supernatant, and resuspend the pellet by tapping the tube (conventional human target leukaemic cell lines to measure NK and LAK activities are K562 [NK-sensitive] and Raji [NK-resistant], respectively).
2. Using a tuberculin syringe with a 23G needle, draw out 0.1 ml ^{51}Cr (Dupont, NEG030, 5 mCi/5 ml) per 1×10^6 target cells to be labelled

(because the half-life of ^{51}Cr is 28 days, you must use twice this amount if ^{51}Cr is $\geqslant$ 4 weeks old). Add the isotope to the cell pellet and incubate at 37 °C for 30–90 min in a CO_2 incubator with periodic tapping (lead-shielding should be used for the radioactive working area and around radioactive waste flasks). Suspend the labelled target in medium and spin at 4°C (450 *g*) for 10 min. Aspirate the supernatant into the radioactive waste flask. Wash two more times and resuspend the cells in complete medium (RPMI or IMDM with 10% FBS) at 2×10^5/ml.

3. While the target cells are being labelled, wash and resuspend the effectors at 10^7/ml in complete medium. Plate them at 200 µl per well in columns 1–3 of a 96-well round-bottom plate (Falcon). Dispense 100 µl of complete medium in columns 4–12; with a multi-channel Titertek pipette (Flow), perform sequential twofold dilutions from columns 1–3 into columns 4–6, then 6–9, and finally 10–12. This results in triplicate wells of four twofold dilutions (from 1×10^6 to 1.25×10^5 effectors/well).
4. Add the labelled targets (10^5 cells/50 µl/well) and incubate the plates for 4 h at 37°C in a 10% CO_2 incubator. Triplicate control wells will contain (a) 50 µl of target and 100 µl of medium (this will give the percentage of spontaneous ^{51}Cr release), and (b) 50 µl target and 100 µl 1% Triton (Sigma) (to measure maximum isotope release).
5. Spin the plates at 400 *g* for 3 min at 4°C. Transfer 100 µl of supernatants into 6 × 50 mm glass tubes (ParaScientific Co.) with a multi-channel Titertek pipette.
6. Measure the sample radioactivity using a gamma counter. Calculate the percentage lysis with the following equation: $100 \times (E\text{–}S/T\text{–}S)$, where E is the isotope release from experimental wells, S is spontaneous release (medium alone), and T is the total release (1% Triton).

Note: some types of targets release high levels of ^{51}Cr spontaneously. Whereas 5–10% spontaneous release is acceptable, values greater than 30–40% render the computation of lytic activity less accurate. For this reason, it is important (at least with these targets) not to incubate the plates longer than 4 h, and to wash the cells properly to ensure that the level of cell-free ^{51}Cr after the last wash is less than 5% of the maximal releasable level. We also recommend the use of target cells in a rapid phase of growth.

5.4.1 Conventional and reverse antibody-dependent cell-mediated cytotoxicity (ADCC)

In a 'conventional' ADCC assay, the ^{51}Cr-labelled target cell is coated with a specific antibody; the effector cell bearing receptors for the Fc portion of

IgG (IgG-FcR) binds to the antibody and kills the target. The 'reverse' ADCC assay is used to test the ability of a given mAb to induce cytotoxic activity in an effector cell against IgG-FcR$^+$ targets (conventionally, the murine mastocytoma cell line P815 or the human monoblastic leukaemia THP-1 are used as targets). Using this assay, we have shown the ability of various mAb that recognize CD2, CD3, CD56, and several other adhesion or activation molecules to induce redirected lysis of P815 and THP-1 targets in our killer T-ALL clones (30).

Protocol 9. Reverse ADCC

1. Dispense the effector cells in plates at appropriate E/T ratios, as in *Protocol 8*.
2. Wash and resuspend the ^{51}Cr-labelled IgG-FcR$^+$ target in medium containing the experimental mAb (ascites, diluted 10^{-2}, or hybridoma supernatant, undiluted). Plate the target/antibody mixture at 50 μl per well on top of the effectors.
3. Incubate the plate at 37°C for 4 h and measure isotope release, as in *Protocol 8*.

5.5 Role of Ca^{2+} in the killing process

Both exocytosis of cytotoxic granules and PFP assembly on the membrane of the target cells have been shown to depend on the presence of extracellular Ca^{2+} (31, 32). To investigate the requirement for extracellular Ca^{2+} for expression of cytotoxic activity against tumour cells, the cytotoxic assays are performed in the presence and absence of the Ca^{2+} chelating agent EGTA.

Protocol 10. Analysis of extracellular Ca^{2+} requirement for lysis

1. Mix the effector and target cells at appropriate E/T ratios in three different medium conditions:
   - complete IMDM (containing 1 mM $CaCl_2$, 0.4 mM $MgSO_4$, and 5% dialysed FBS)
   - complete IMDM with added 2 mM [ethylene bis(oxyethylenenitrilo)]-tetraacetic acid (EGTA) and 2 mM $MgCl_2$ (the excess of $MgCl_2$ guarantees the presence of free Mg^{2+}, which is required for E/T binding)
   - complete IMDM with added 2 mM EGTA, 2 mM $MgCl_2$, and 2 mM $CaCl_2$

   As a control for EGTA toxicity, always incubate effectors and targets separately in 4 mM EGTA overnight and monitor their viability the next day.

2. Dispense the E/T mixtures set up in the three different medium conditions in microtitre plates.
3. Perform a 4 h-^{51}Cr release assay, as described in *Protocols 8* and *9*.

In our experience, the Ca^{2+} chelating agent EGTA can be used without toxic effects at a concentration of 4 mM. As an example, the spontaneous killer activity of two T-ALL cell lines established in this laboratory is totally abrogated in the presence of 2 mM EGTA, but can be efficiently restored by adding an excess of Ca^{2+} (4 mM). This indicates that the loss of cytotoxicity is due to the Ca^{2+} chelating action rather than to a direct inhibitory activity of EGTA (30). On the other hand, the requirement for extracellular Ca^{2+} by the T-ALL cell lines in reverse ADCC assays is only partial (30), demonstrating that mAb redirected lysis occurs through mechanisms mostly independent of exocytosis of cytolytic granules.

5.6 Exocytosis of cytotoxic granules

Serine esterase 1 (SE1) release into the supernatant has been used as a convenient indicator for degranulation by CTL and LAK cells (31, 33). To investigate whether granule exocytosis plays a role in the lytic mechanism of a given effector population, BLT-esterase secretion can be studied in response to stimulation by mAb (in reverse ADCC assays) or exposure to tumour targets (direct lysis).

SE1 release is quantitated by spectrophotometric measurement of the coloured product of the enzymatic degradation of a synthetic substrate (BLT) (33). Because SE and PFP are confined to the same granules, both in NK cells and CTL, BLT-esterase secretion can be considered as a marker for PFP exocytosis as well.

Protocol 11. BLT-esterase release assay upon activation with antibodies

1. When the activating mAb (ascites, at 10^{-2}, or undiluted supernatants) is used in a soluble form, incubate the effector cell with the mAb at 29°C for 30 min. If immobilization of the mAb is preferred, dispense the mAb (diluted 1:1 in $NaCO_3/NaHCO_3$ 0.1 M, pH 9.6) into a 96-well round-bottom microtitre plate (50 µl/well), and incubate overnight at 4°C (put 50 µl of PBS in the control wells). Then wash the plate with RPMI-1640 medium supplemented with 5% FBS and 10 mM Hepes buffer.
2. Distribute 5 × 10^5 effectors and 1 × 10^5 targets/well, and incubate for 4 h at 37°C in a 96-well microplate (30). (Longer incubation times are often accompanied by increased background enzyme release.)

Protocol 11. *Continued*

3. Centrifuge the plate for 5 min at 400 *g* and transfer 175 μl of culture supernatant to another 96-well microplate, called the 'supernatant' plate. (This plate may be frozen at −20 °C if a later read-out of the assay is desired.)
4. Remove the left over 25 μl and resuspend the pellets carefully in 200 μl of medium. Prepare cell lysates from these pellets by four cycles of freezing and thawing. (Use a cardboard box filled with solid CO_2 for freezing; for thawing, put the plate in a water bath at 99 °C over a sponge.) Transfer the content of the wells into a second 96-well microplate called the 'cell lysate' plate.
5. Prepare 100 ml of BLT substrate solution immediately before developing the reaction, by adding the following reagents to a conical plastic tube:
   (a) 100 ml 0.1 M Tris–HCl, pH 8.0
   (b) 10 mg BLT stock (2×10^{-4} M *N*-benzyloxycarbonyl-L-lysine thiobenzyl ester) (Calbiochem)
   (c) 10 mg DTNB stock [1.1×10^{-4} M 5,5′dithiobis-(2-nitrobenzoic acid)] (Pierce)
6. Transfer 50 μl of supernatant and cell lysate samples into a third 'assay' plate and add 150 μl of the substrate. The reaction is allowed to develop for 30 min at room temperature, and the absorbance is measured using a 412 nm wavelength filter in an ELISA plate reader.
7. Data are presented as specific percentages of enzymatic activity released, based on the equation $100 \times (E - S/T - S)$, where E represents the number of enzymatic units in the supernatants of experimental wells, S is the spontaneous BLT-esterase release from unstimulated wells, and T is the total release from the cell lysates.

Note: when tumour target cells rather than antibodies are used as stimulus for lymphocyte activation and release of SE1, effectors and targets (both 1×10^5 well) are incubated together for 4 h at 37 °C; the rest of the BLT-esterase assay is the same.

Protocol 12. Detection of BLT-specific proteases at the single cell level

This staining technique allows the visualization of SE1 in the intracytoplasmic granules of cytotoxic cells (34).

1. Activate 10^5 effector cells by incubation with targets or mAbs (see *Protocol 11*) and concentrate the cells on slides by cyto-centrifugation. Fix-dry the slides by immersion in cold acetone for 10 min at 4 °C.

2. Incubate the slides at 37°C for 15 min with 2×10^{-4} M BLT substrate in 0.2 M Tris–HCl buffer, pH 8.1, in the presence of 0.2 mg/ml incorporating Fast Blue BB salt (Sigma) as a capture chromogenic agent.
3. Rinse the excess reaction mixture from the slides by immersion in water for 1 min, and counterstain the slides in haematoxylin for 5 min.
4. Rinse, dry, and mount the slides under a coverslip with 50% glycerol in PBS, and examine them immediately under a microscope.

5.7 Proliferative responses

Another way to study lymphocyte activation is to measure their proliferative responses to a given stimulus in a short-term ^{3}H-TdR incorporation assay. An example of proliferation in response to mAb is provided below.

(i) Pre-coat 96-well microplates with the mAb in carbonate buffer, pH 9.6, and incubate the plates overnight at 4°C.

(ii) Seed 5×10^4/well effector cells and incubate for 3–4 days at 37°C. Add ^{3}H-TdR (1 μCi/well) 6–18 h later. Proceed as in *Protocol 2*.

5.8 Lymphokine production

Activated lymphocytes are able to produce lymphokines, such as interferon and tumour necrosis factor, which play an important modulatory role in inflammation and immune responses (35, 36). We have shown that our cytotoxic T-ALL cell lines are able to release, rapidly and efficiently, high levels of interferon-γ, tumour necrosis factor-α, GM-CSF, transforming growth factor-β1 and IL-8 upon induction with cytokines (IL-2 or IL-12), certain mAbs and susceptible tumour targets (19, 28, 37, 38). To determine the ability of tumour cells or antibodies to trigger lymphokine release in human lymphocytes, we suggest the following procedure.

(i) Incubate the lymphocytes at 37°C in a humidified CO_2 incubator for 16–20 h in the presence of the inducer (0.2×10^6/ml target cells or ascites at 10^{-2}).

(ii) After centrifugation at 4°C at 450 *g* for 10 min, harvest the cell-free supernatant, filter it on a 0.2 μm filter (Costar), and perform a radioimmunoassay or ELISA, according to described techniques (37, 38).

Acknowledgements

This work was supported by NIH grants CA-47589 and CA-10815, ACS grant CH-527, and the José Carreras International Leukemia Foundation.

References

1. Minowada, J. (1982). In *Leukemia* (ed. F. Gunz and E. Henderson), pp. 119–25. Grune and Stratton, Orlando, FL.
2. Poiesz, B. J., Ruscetti, F. W., Mier, J. W., Woods, A. M., and Gallo, R. (1980). *Proc. Natl. Acad. Sci. USA*, **77**, 6815.
3. Moore, M. A. S., Williams, N., and Metcalf, D. (1973). *J. Natl. Cancer Inst.*, **50**, 603.
4. Griffin, J. D. and Lowenberg, B. (1986). *Blood*, **68**, 1185.
5. Nara, N. and McCulloch, E. A. (1985). *Blood*, **165**, 1484.
6. Dicke, K. A., Spitzer, G., and Aherarn, M. J. (1976). *Nature*, **295**, 129.
7. Griffin, J. D., Young, D., Hermann, F., Wiper, D., Wagner, K., and Sabbath, K. D. (1986). *Blood*, **67**, 1448.
8. Hoang, T., Nara, N., Wong, G., Clark, S., Minden, M. D., and McCulloch, E. A. (1986). *Blood*, **68**, 313.
9. Lange, B., Valtieri, M., Santoli, D., Caracciolo, D., Mavilio, F., Gemperlein, I., Griffin, C., Emmanuel, B., Finan, J., Nowell, P., and Rovera, G. (1987). *Blood*, **70**, 192.
10. Santoli, D., O'Connor, R., Cesano, A., Phillips, P., Colt, T. L., Lange, B., Clark, S. C., and Rovera, G. (1990). *J. Immunol.*, **144**, 4703.
11. O'Connor, R., Cesano, A., Lange, B., Finan, J., Nowell, P. C., Clark, S. C., Raimondi, S. C., Rovera, G., and Santoli, D. (1991). *Blood*, **77**, 1534.
12. Griffin, J. D. and Lowenberg, B. (1986). *Blood*, **68**, 1185.
13. Ozawa, K., Hashimoto, Y., Urabe, A., Suda, T., Motoyoshi, K., Takaku, F., and Miura, Y. (1982). *Exp. Hematol.*, **10**, 145.
14. Metcalf, D. (1984). *Clonal cultures of hemopoietic cells: techniques and applications*. Elsevier, Amsterdam.
15. Bosma, G. C., Custer, R. P., and Bosma, M. J. (1983). *Nature*, **301**, 527.
16. McCune, J. M. (1991). *Curr. Opin. Immunol.*, **3**, 224.
17. Kamel-Reid, S., Latarte, M., Sirard, C., Doedeus, M., Grunberger, T., Fulop, G., Freedman, M. H., Phillips, R. A., and Dick, J. E. (1989). *Science*, **246**, 1597.
18. Cesano, A., O'Connor, R., Lange, B., Finan, J., Rovera, G., and Santoli, D. (1991). *Blood*, **77**, 2463.
19. Cesano, A., Hoxie, J. A., Lange, B., Nowell, P. C., Bishop, J., and Santoli, D. (1992). *Oncogene*, **7**, 827.
20. Hendrickson, E. A. (1993). *Am. J. Pathol.*, **143**, 1511.
21. Kyoizumi, S., Baum, C. M., Keneshima, H., McCune, J., Yee, E. J., and Namikawa, R. (1992). *Blood*, **79**, 1704.
22. Namikawa, R., Ueda, R., and Kyoizumi, S. (1993). *Blood*, **82**, 2526.
23. Berke, G. (1991). *Curr. Opin. Immunol.*, **3**, 320.
24. Ortaldo, J. R. and Hiserodt, J. C. (1989). *Curr. Opin. Immunol.*, **2**, 39.
25. Young, J. D. E., Podack, E. R., and Cohn, Z. A. (1986). *J. Exp. Med.*, **164**, 144.
26. Young, J. D. E., Liu, C. C., and Persechini, P. M. (1988). *Immunol. Rev.*, **103**, 161.
27. Taiu, Q., Strenli, M., Saito, H., Schlossman, S. F., and Anderson, P. (1991). *Cell*, **67**, 629.
28. Cesano, A., Visonneau, S., Clark, S. C., and Santoli, D. (1993). *J. Immunol.*, **151**, 2943.

29. Callewaert, D., Radcliff, G., Waite, R., LeFevre, J., and Poulik, M. D. (1991). *Cytometry*, **12**, 666.
30. Cesano, A. and Santoli, D. (1992). *In Vitro Cell. Dev. Biol.*, **28A**, 648.
31. Ostergaard, H. and Clark, W. R. (1987). *J. Immunol.*, **139**, 3573.
32. Engelhard, V. H., Gnarra, J. R., Sullivan, J., Mandell, G. L., and Gray, I. S. (1988). *Ann. NY Acad. Sci.*, **142**, 4378.
33. Takayama, H., Trenn, G., and Sitkovsky, M. V. (1987). *J. Immunol. Methods*, **104**, 183.
34. Wagner, L. W. B., Wiescholzer, V., Sexl, H., Bognar, H., and Worman, C. P. (1991). *J. Immunol. Methods*, **142**, 147.
35. Peters, P. N., Ortaldo, J. R., and Shalaby, M. R. (1986). *J. Immunol.*, **137**, 2592.
36. Cuturi, M. C., Murphy, M., and Costa-Gianni, M. P. (1987). *J. Exp. Med.*, **165**, 1581.
37. Cesano, A. and Santoli, D. (1992). *In Vitro Cell. Dev. Biol.*, **28A**, 657.
38. Visonneau, S., Cesano, A., and Santoli, D. (1994). In *Leukocyte Typing V* (ed. S. F. Schlossman), Oxford University Press, Oxford. (in press).

12

Senescence and immortalization of human cells

KAREN HUBBARD and HARVEY L. OZER

1. Introduction

Normal vertebrate diploid cells in culture have a finite lifespan (1). The maximum number of population doublings is dependent on cell type and species. It has also been shown to vary as a function of age of the donor organism, resulting in its utilization as a model for the proliferative aspects of cellular ageing. Although cells from all species eventually cease to proliferate and become 'senescent', the capability to achieve infinite or continuous cell growth (i.e. became 'immortal') by mechanisms which overcome cellular senescence is strikingly dependent on the animal species. Rodent cells readily become 'spontaneously' immortal such that there are multiple established fibroblastoid cell lines; for example, mouse 3T3 cells. In marked contrast normal chicken and human fibroblasts (HF) rarely, if ever, become immortal. The observation that many human tumours are immortal cell lines when introduced into tissue culture suggests that overcoming senescence is a stage in carcinogenesis. One experimental manipulation which has been successful in facilitating the immortalization of HF has been through the introduction of genes from DNA tumour viruses into otherwise normal cells. In particular, the gene encoding the large T antigen (or A gene) from SV40 has been especially useful in this regard. Microinjection of large T antigen will reactivate DNA synthesis in senescent cells, although mitosis is not induced. Stable introduction of the A gene (as by viral infection or DNA-mediated gene transfer) prior to senescence will extend the limited lifespan for multiple generations; however, such cells typically undergo cell death, termed 'crisis' as distinct from senescence. These SV40 transformed cells can also be considered pre-immortal cells since they have an increased likelihood of becoming immortal, albeit with a varied and, most often, low frequency. Indeed many continuous cell lines have been developed *in vitro*. It should be emphasized that SV40-immortalized cells are still dependent on large T antigen function. We (2) and others (3) have developed a two-stage model. In the first stage large T antigen functions as a mitogenic agent and transiently overcomes the basis for cellular senescence. A cellular mutation is required for the second stage. A gene has been localized to the long arm of chromosome 6, whose inactivation is consistent with being responsible for this step (4). The remainder of the chapter describes each of these stages of cellular proliferation; senescence, crisis, and immortalization. In addition, we will summarize how SV40 and, especially its *tsA* mutants encoding a heat-labile large T antigen, have been used to develop cell lines for multiple lines of experimentation.

2. Cellular senescence

Primary cells introduced into culture often proliferate when given appropriate medium and other culture conditions. Such media have been developed for multiple cell types and include examples of fully defined media as well as those supplemented with animal sera, most often bovine sera, as discussed elsewhere (Chapter 13). This period may include an initial interval of adaptation but is often followed by a stage of repeated and sometimes extensive cell proliferation. In the case of human fibroblasts derived from fetal tissue, it may exceed 50 population doublings (PD) (and an approximately equal number of cell generations). This is followed by a plateau period during which the population appears to slow down due to an increase in generation time and a decreased likelihood of cell division, resulting finally in a cessation of cell proliferation or senescence.

Loss of cellular proliferation in senescence is a stochastic process at the individual cell level. The mechanism is the subject of some controversy (5). Earlier models suggesting an accumulation of mutations in nuclear DNA or other errors have become less favoured because of the absence of strong supporting data although there has been recent evidence of alteration in mitochondrial DNA as a contributing factor. On the other hand, senescence does share features consistent with replicative inhibition associated with terminal differentiation. In the case of human fibroblasts, it would appear to be at least an important experimental model for cellular ageing, regardless of the underlying mechanism. Cellular proliferative capacity exhibits a range of values as determined by size and number of single-cell derived colonies at low cell density, consistent with the behaviour of the population at higher cell density. Similarly assessment of DNA synthesis by [^{3}H]thymidine autoradiography indicates a heterogeneous population of cells with regard to generation time and number of replicatively active cells. This heterogeneity decreases with the proliferative history of culture. Both the percentage of cells undergoing DNA synthesis (*Protocol 1B*) and colony formation can therefore be used to assess a particular point in time in the lifespan of a population of cells in comparison to a reference cell line and the same cell line early in its lifespan. More recently, it has been recognized that somatic cells undergo a shortening of the repetitive sequences at the telomeric ends of chromosomes. This is due to the special requirements for replication of DNA termini and several factors including a specific replicase called telomerase. There is no detectable telomerase activity in HF (and many other cell types) such that HF lose approximately 40 bp per cell generation. Senescent cells have considerably shortened telomeric sequences. Therefore the length of telomeric sequences by Southern blot analysis after digestion of appropriate restriction enzymes may be useful in determining the replicative capacity of the cell population. It should be noted that models based on loss of telomeric sequences have been proposed to explain senescence (6); however, such cells can undergo further shortening as will be discussed later.

In any case, when fetal human fibroblasts are studied, remarkable similarity is observed with the total lifespan of 60–70 PD not only for a given cell line within a laboratory but among cell lines (WI–38, IMR90, HS74BM, TIG) in different laboratories when subculture ratios of 1:3 to 1:5 in conventional growth medium containing 10–15% fetal bovine serum are employed. None the less it is necessary and appropriate for a particular laboratory to determine lifespan by serial passage. Senescence is defined by the absence of one PD in 2–4 weeks at terminal passage. Cells assume a typical morphology characterized by marked cellular enlargement and a flattened appearance. Less than 5% of the cells undergo DNA synthesis in a 48 h period, shown by the incorporation of [^{3}H]thymidine into nuclear DNA as determined by autoradiography. Cells have a G_1 content of DNA, although it should be recognized that tetraploidy and aneuploidy commonly occur late in the fibroblast lifespan. The determination of lifespan and methodology for assessing whether a culture has become senescent are described in *Protocol 1*.

Protocol 1. Determination of lifespan and preparation of senescent cell populations

A. *Cell cultures*

- Human fibroblast and other cell lines of limited lifespan at determined PD can be obtained from the National Institute of Aging Collection in the Cell Culture Repository at the Coriell Institute, Camden, NJ.
- WI-38 cells may be obtained from V. Cristofalo, Center for Gerontological Research, Medical College of Pennsylvania, Philadelphia, PA.
- Procedures and reagents for cell culture are described in Chapter 13 and elsewhere in the volume.

1. For estimation of lifespan, subculture human fibroblasts at 1:5 ratios in growth medium containing 10% fetal bovine serum (FBS).
2. Determine cell numbers either electronically using a Coulter Counter or manually with a haemocytometer.
3. Calculate the number of population doublings achieved using the

Protocol 1. *Continued*

following formula: $N_C/N_S = 2X$ *(where* C = number cells harvested at confluency; S = initial number cells seeded; X = population doublings). Cells are considered senescent when the labelling index of nuclei and the rate of DNA synthesis is less than 5% of that found for early passage cells.

Once the population doubling for senescent cells has been determined, this information can be used for larger scale culture preparations. Frozen cultures at intermediate passage levels represent useful starting cultures.

B. *Labelling index of nuclei*

1. At each subcultivation to be tested, seed cells at $1–5 \times 10^4$ cells/35 mm tissue culture dish.
2. After 24 h, add [^{3}H]thymidine at a final concentration of 0.1–1 μCi/ml.
3. Wash cells at time intervals of 1 to 3 days and process for autoradiography (all operations are on ice).
4. Wash cells 2× with cold phosphate-buffered saline (PBS) followed by a 15 min incubation with 5 ml cold methanol.
5. Remove methanol and incubate cells with 5 ml cold 5% trichloroacetic acid (TCA) for 30 min (this allows for extraction of unincorporated radioactivity).
6. Wash cells 2× with distilled H_2O and allow to air dry.
7. For autoradiography, add 0.5 ml pre-warmed Kodak emulsion (50°C) to each dish followed by gentle rocking to ensure entire coverage of surface (all operations are done in the dark).
8. Pour excess emulsion off the dish. Cover dishes with foil and leave in the dark for 1 week.
9. Add 1:1 dilution of Kodak D-19 developer (pre-warmed to 59°C) to each dish and incubate for 4 min at room temperature.
10. Wash dishes with distilled H_2O.
11. Add Kodak acid fixer and incubate dishes for 5 min. Pour off the fixer and add distilled water to each dish.
12. Allow dishes to air dry and count the number of labelled nuclei using an inverted microscope. (Staining of cells facilitates detection of unlabelled cells.)

C. *Rate of DNA synthesis*

1. Seed 1×10^5 cells per 100 mm dish 24 h prior to [^{3}H]thymidine incubation.
2. Wash cells at time intervals of 1 to 3 days with cold PBS.

3. Harvest cells by trypsinization and dilute with PBS. Take an aliquot for cell counting; centrifuge the remainder and resuspend in 1 ml cold PBS by vortexing.
4. Add 1 ml of 2× lysis buffer (0.2% SDS, 20 mM Tris–HCl, pH 7.5, and 2 mM EDTA); vortex the tube and incubate at room temperature for 10 min.
5. Add 2 ml of 20% TCA and store the sample on ice for 1 h.
6. Filter precipitated DNA through pre-wet GF/A filters (Whatman) and wash with 5% TCA (3 times) followed by cold 95% ETOH.
7. Count filters in a scintillation cocktail. (Sensitivity can be improved by solubilization of samples prior to the addition of scintillant.)

Although the passage at which the population is determined to be senescent is often quite reproducible under a standard set of culture conditions, it may vary by one or two passages, thus complicating experimental designs utilizing large-scale cultures. Many laboratories have therefore chosen to employ an alternate procedure (*Protocol 2*). Under these conditions cells are growth arrested, taking advantage of the fact that normal fibroblasts show density-dependent inhibition of growth. Cells are cultured until they reach confluence and further growth arrested by incubation in serum-free (or low, 0.5%, serum) for 3–4 days. Such cultures are arrested in G_0 with less than 5% of the population undergoing DNA synthesis. Such 'quiescent' cultures prepared from cells which have completed 90–95% of their calculated lifespan do not efficiently resume DNA synthesis when the medium is replaced with conventional growth medium supplemented with 10–20% fetal bovine serum or purified growth factors (see Chapter 13) whereas quiescent cultures of early passage cells typically show 40–70% of the cells entering S phase under the same condition. The late passage cultures closely mimic senescent cultures which have been similarly treated with fresh media with serum.

Protocol 2. Preparation of large scale cultures of senescent cells through G_0 arrest

1. Subculture cells at 1:5 ratios until they have reached 95% of their lifespan (as predicted by smaller scale determination as described in *Protocol 1*).
2. Expand cultures to produce the desired number of cells by seeding at 5×10^5 cells per 100 mm dish.
3. Allow cells to reach confluence. At this point, cells should be near 99% completion of their lifespan. Remove media, wash twice with serum-free medium, then re-feed with medium containing 0.5% FBS. Incubate for 3–4 days.

Protocol 2. *Continued*

4. Restimulate cells with medium containing 10% FBS, harvest after 24 h and use cells in appropriate assays. (To ensure that these cells do not re-initiate DNA synthesis, replicate dishes can be assessed as described above in *Protocol 1B* or *1C* using cells at earlier passage as control.)

The major advantage of the method in *Protocol 2* is its convenience. The major disadvantage is that data from some laboratories suggest that the two experimental systems are not equivalent. For example, cells in the quiescent cultures still retain the potential for further cell division and may therefore retain relevant biochemical functions (at unknown levels) which are virtually absent in senescent cells. This difficulty can be avoided in part by comparing a specific property in quiescent late passage cultures with that in quiescent senescent cultures. Quiescent cultures of cells early in their proliferative lifespan, i.e. less than 50% of the total lifespan, also need to be evaluated to verify that the property is associated with senescence rather than a characteristic of quiescent cultures, in general. Indeed, assessment of cells at different cell passage is strongly advised. It should be noted that senescent cells are likely not to be equivalent to G_0 quiescent cells. It has been proposed that senescent cells are arrested later in G_1 (near the onset of S phase) than quiescent cells which are commonly considered to be in early G_1, based on the elevated thymidine triphosphate pools in the former. In any case, it is clear that serum stimulation of both senescent and late passage quiescent cultures fails to induce DNA synthesis and that such treated cultures share many biochemical properties. These features have been used to help define the block in proliferation associated with senescence.

Recently, it has been shown that senescent cells are defective in expression or function of several important growth-regulatory molecules associated with progression of cells through G_1 into S. It was first reported that the induction of c-*fos* and the activity of its transcription factor form AP-1 (composed of c-*fos* and *jun*) were decreased in senescent fibroblasts. However, microinjection of purified c-*fos* or its cDNA in a suitable expression vector did not in itself induce DNA synthesis. More recently, it was found that the growth suppressor pRb-1 is aberrantly regulated. In G_0 or early G_1 the pRb-1 protein is hypophosphorylated and becomes phosphorylated as cells progress through G_1 into S. In this manner the association between pRb-1 and the transcription factor E2F-1 becomes altered. E2F-1 stimulates transcription of several genes involved in DNA synthesis and although multiple other proteins may be involved, it is clear that the phosphorylation of pRb-1 and release of E2F-1 is a key step in the regulation of E2F-1. Most interestingly it has been shown that pRb-1 is underphosphorylated in senescent cells, even after such cells are stimulated to express other factors by treatment with serum-derived growth factors (7). Cellular kinases (cyclin-dependent kinases, CDK) have been identified which may be responsible. Most recently a new class of inhibitors of such CDK have been identified and cloned. One such member, variously called p21, CIP, WAF, and sdi-1, has been shown to be over-expressed in senescent and quiescent cells. Furthermore, it inhibits cell proliferation when expressed at high levels in growing cells. Interestingly, its gene expression is induced by the growth suppressor p53. Taken together, these results support the model that cellular senescence is dependent on normal growth regulatory factors. The appendix to this chapter lists a variety of properties which have been reported to be different in senescent cells.

Studies with SV40 large T antigen, which will be discussed in more detail in subsequent sections, further support this interpretation. Large T antigen has multiple effects on cellular gene expression in different experimental systems. It has direct as well as indirect effects on viral and cellular DNA synthesis. This protein can induce cellular DNA synthesis in both quiescent and senescent cells. It is capable of forming complexes with pRb-1 and p53. When large T complexes with pRb-1, E2F-1 is released from pRb-1. Mutants of large T antigen which cannot bind pRb-1 do not induce cell DNA synthesis. Furthermore, the subunit composition of cyclin–CDK complexes is rearranged in SV40-transformed cells. These changes may either be direct effects of large T antigen or mediated through inhibition of p53 activated transcription of the CDK inhibitor WAF.

3. SV40-transformation and crisis

Introduction of an SV40 genome into multiple cell types extends the proliferative lifespan of such cells and increases the likelihood that such cells will become immortal. Although this has been demonstrated in rodent cells as well as human cells, this approach has been most useful in studies with human cells. It is relatively simple to obtain cultures or cloned cell lines expressing an SV40 genome since most human cells are susceptible to infection with SV40 virus particles or can be transfected by purified DNA. The SV40 genome is 5.2 kb. Its sequence is known and multiple mutations in all of its functions are available. Its genome can be divided into three regions; a control region (C), the region expressed early in infection (E), and the region expressed late in infection (L). The control region contains the sequence required for DNA replication (origin) flanked by the promoter and the enhancer for the E region as well as the promoter for the L region. The E region is required for transformation and DNA synthesis. It encodes two polypeptides: large T antigen (100 kDa) and small t antigen (21 kDa). The L region encodes the viral structural proteins and is not involved in SV40-mediated transformation. The course of SV40 infection is species dependent. It replicates efficiently in several monkey cell lines, especially CV-1, BSC-1 or VERO cells. Infection of rodent cells is considered non-permissive as it does not produce progeny virus. Replication of its DNA is very low or undetectable in mouse or rat established cell lines; it is markedly reduced (~99%) in Chinese hamster cell lines. SV40 infection of human cells is semi-permissive. Low but readily detectable levels of progeny virus are produced (typically 10^6 infectious units per culture of 10^6 cells). However, viral DNA synthesis is quite efficient in many cells with as many as 10 000 copies per cell reported. Infection of human fibroblasts by SV40 virus results in a mixed cell population. Some cells are productively infected producing high levels of DNA and virus, and die within a few days due to the cytopathic effect of the virus. Most cultures establish a persistent infection with stable expression of a low level of virus production. The viral genome is integrated into many cells resulting in transformation. However these cells are capable of excision and replication of the viral genome. Finally some cells are not infected even at high multiplicity of infection but can be subsequently infected by progeny virus. Virus spread can be minimized by including antibody to neutralize SV40 virus particles in the culture medium.

Non-productive SV40 infection of permissive and human cells can be obtained by introduction of defective viral genomes which cannot produce virus particles due to deletion or disruption of L region coding sequences. This commonly occurs as part of the cloning process in generating a recombinant DNA in a plasmid vector. Such constructs can still replicate DNA efficiently in human cells but are not expected to produce extracellular virus. (If the L region is only interrupted, recombinational mechanisms may result in an intact viral genome.) We have emphasized in this laboratory the use of cloned origin-defective mutants of SV40, especially those in which the *Bgl*1 site has been deleted due to removal of a few basepairs. These ori-minus constructs were prepared by Y. Gluzman. They express both SV40 T antigens at normal levels but do not replicate DNA or produce progeny virus. They integrate stably into human fibroblasts at 1–2 copies per cell. Integrants have been selected based on the transformed phenotype or using a co-selectible marker for resistance to the neomycin analogue G418. Constructs in which the SV40 origin has been deleted altogether provide a similar benefit. Such examples utilize the long terminal repeat (LTR) regulatory sequences from Rous Sarcoma virus (RSV) or murine leukaemia virus (MLV), the glucocorticoid-responsive LTR from mouse mammary tumour virus (MMTV), or the cellular metallothionein promoter. It should be noted that the advantage would be negated by the inclusion of drug-resistant sequence containing an SV40 origin (such as SVneo, for example) in the DNA construct.

It was recognized early that SV40-transformed human cells were not typically immortal. In the original studies, mixed cultures of transformed and non-transformed cells were passaged by conventional subcultivation until all the cells were positive for large T antigen by immunofluorescence. However, after multiple successful passages, the culture underwent a series of morphological changes with marked cell rounding and cell death, termed crisis—'a period of balanced cell growth and cell death followed by a decrease in the total number of surviving cells' (8). Our laboratory has isolated a series of independent SV40-transformed human fibroblasts (SV/HF) generated with cloned ori-minus SV40 DNA (2, 4, 9). All transformants were 100% large T antigen-positive upon isolation (as a transformed focus or colony in agarose). All showed an extended lifespan, 80–90 PD versus the parental HF which senesced at 64 PD. However, all underwent crisis despite the absence of viral DNA synthesis, ruling out virus production or viral-mediated cell death being responsible for crisis. Although the mechanisms of extended lifespan and crisis are still not understood in detail, several aspects are evident. First, extension of lifespan requires expression of large T antigen but does not require small t antigen, since we have found that constructs in which a small deletion affecting small t antigen but not large T antigen still show extended lifespan. Such transformants do not show classical transformed foci; the transfectants were isolated by including the gene for resistance to G418 using a non-SV40 promoter (i.e. RSVneo) in the vector. Secondly, persistent growth requires continuous large T antigen expression. HF transformed by origin-defective SV40 encoding a heat-labile, temperature-sensitive (ts) large T antigen (SVtsA58) display extended lifespan when cultured at 35°C, the permissive temperature for large T function. When cultures of such transformed cells (SVtsA/HF) are shifted to 39°C, large T function is lost and the cells undergo growth arrest, exhibiting morphological changes reminiscent of senescence. Thirdly, large T functions involved in inactivation of pRb-1 and p53 are likely to be involved. Large T forms complexes with both proteins at 35°C; however, no complexes are observed at 39°C with either protein. Immunoreactive large T antigen is still present although at reduced levels. Fourthly, it has also been demonstrated (10) that the large T–p53 complex is important since HF infected by a mutant SV40 which is unable to bind human p53 does not have the extension in lifespan. In view of the role of pRb-1 and p53-induced gene products in senescence, it is not surprising that such functions are important to SV40-mediated extension of lifespan as well.

A major unanswered question remains why the initial ability of large T antigen to extend the lifespan ceases to be sufficient. Although no answers are currently available, two observations are relevant. SV/HF show an increased requirement for growth factors, suggesting a decreased responsiveness to them. Although the mechanism is unknown, it is known that reducing the concentration of growth factors can induce an injury response pathway and trigger apoptosis. Indeed several features of crisis are similar to apoptosis and unpublished data obtained in this laboratory show internucleosomal DNA fragmentation, typical of apoptosis, in SV/HF at crisis. Also, shortening in cell telomeric DNA sequences continues during this period of extended lifespan. Therefore, telomeres may have reached a critical length that results in major changes in chromosome stability or chromatin structure. Alternatively, the persistent inactivation of p53 (through complex formation with large T) and the loss of normal checkpoint controls may have accumulated sufficient DNA damage that cells are no longer viable. A limited number of studies have examined gene expression in cells during or approaching crisis. Several mRNAs characteristic of senescent cells are elevated including the CDK-inhibitor (sdi-1); however, the significance of these changes is unclear.

4. SV40-immortalized cell lines

A rare SV40-transformed human cell becomes immortal. Since all SV/HF cells express large T antigen and there is no evidence for major changes in SV40 function in the immortal as compared to the pre-immortal SV/HF (3), attention has focused on non-SV40 functions. Several lines of evidence support this approach. First, immortalization is recessive to limited lifespan in cell hybrids. Hybrids between senescent cells and normal cells have reduced lifespan consistent with the presence of an inhibitor of cell proliferation in senescent cells. Hybrids between immortal and normal cells indicate that immortal cells are not resistant to such inhibition since the lifespan is limited, consistent with that of the normal cell. Rather the data support the model that immortal cells have inactivated their own senescence gene. Second, hybrids between immortal cells are often not immortal. Pereira-Smith and Smith (11) have shown that hybrids between independent immortal cell lines behave in a fashion consistent with complementation of loss of function. Although all but one of the SV/HF that they tested did not complement each other, multiple cell lines did complement SV/HF and each other. The data fit a minimum of four complementation groups, with SV/HF and some human tumour cell lines in group A. HeLa and several other tumour lines were in group B. It should be noted that HeLa cells are from a patient with cervical carcinoma infected by human papillomavirus (HPV). HPV is related to SV40; both have proteins which bind to pRb-1 and p53 and inactivate their functions, although by different mechanisms. The fact that SV/HF and HeLa are immortal for different reasons further emphasizes that although these are growth suppressors which are pivotal to extended lifespan, they are insufficient for immortalization.

As a prelude to identifying the gene involved in immortalization, several laboratories have employed genetic approaches which have been useful in identifying genes critical in carcinogenesis. We (4, 9) and others (12) have found that loss of sequences on the long arm of chromosome 6 is associated with immortalization of matched sets of preimmortal and immortal SV/HF. The model is that the grossly intact chromosome has a mutated copy of the gene responsible for senescence (*SEN6*) and the other copy is deleted on the rearranged chromosome 6. Hence, there is no functional copy of *SEN6*. Consistent with this model, introduction of a normal chromosome 6 by microcell-mediated cell fusion experiments (MMCT) results in restoration of senescence and the inhibition of cellular proliferation (13). Although immortal tumour cell lines typically have multiple chromosome rearrangements which complicate the first approach, MMCT can be used. Such experiments have been successfully employed by others to demonstrate a putative senescence gene for complementation group B cells on chromosome 4 (14), for group C cells on chromosome 1 (15), and for group D cells on chromosome 7 (16). Furthermore, data thus far also show that introduction of these chromosomes does not suppress growth of other tested complementation groups, ruling out that they are general growth suppressors. It should be noted that the studies do not preclude the presence of multiple growth suppressors on each chromosome being tested. These data explain, at least in part, the observation that immortalization has a virtually undetectable frequency in normal fibroblasts. Since multiple growth suppressors have to be inactivated in human fibroblasts by SV40, multiple mutational events would presumably be required for HF when SV40 is not involved. Indeed, the immortal human fibroblasts used in the studies on chromosome 7 had undergone over 20 treatments with mutagen. Fibroblasts from patients with Li-Fraumeni syndrome have a mutant p53 and have been reported to 'spontaneously' become immortal (17). Such cell lines are highly aneuploid, consistent with loss of p53 checkpoint function and accumulation of genomic rearrangements.

This interpretation also explains why even SV40 transformed HF do not readily become immortal. Indeed studies in this laboratory and others in which pSV3neo transformed HF were studied, the frequency of immortalization for individual transformants differed considerably from one another. About 40% of the individual SV/HF yielded immortalized derivatives which compares favourably with earlier studies using virus-transformed HF. The frequency of immortalized derivatives differed widely, ranging from 10^{-3} to 10^{-7} per cell for different independent transformants. Furthermore 60% of the SV/HF clones did not yield any immortalized derivatives even though all the cells from those colonies also expressed large T antigen and were indistinguishable in growth properties from those that produced immortal cells. This range among individual SV/HF clones suggests that the pre-immortal SV/HF may have undergone a mutation in the *SEN6* gene at different times of its passage history, resulting in variable numbers of heterozygotic cells when the second allele becomes inactivated. The results also suggest that uncontrolled factors may be influencing the outgrowth of immortalized derivatives. *Protocol 3* provides a method for isolation of immortal cell lines. It recommends the use of pooled SV/HF to maximize the likelihood that an SV/HF which has undergone a *SEN6* mutation is present in the population and therefore that an immortalized cell line will be obtained.

Protocol 3. Isolation of SV40 immortal human cells

This method is applicable for transformation by SV40 for many different cell types. Normal fetal HF characteristically senesce at 60–65 PD and lifespan extension brings it up to approximately 90 PD, therefore, immortal derivatives are defined to exceed at least 100 population doublings.

Protocol 3. *Continued*

1. Subculture adherent cells in suitable serum containing medium.
2. Plate 5×10^5 cells per 100 mm dish at 37°C, 24 h prior to transfection of DNA.
3. Transfection of SV40 DNA can be done using several methods. This laboratory routinely uses the calcium phosphate (Ca-P) DNA co-precipitation method. 1–10 μg SV40 DNA (ori-minus) in plasmid vectors is used plus 10–20 μg calf thymus DNA per dish. Mock transfections utilize only calf thymus DNA as control. All subsequent steps and subcultures of cells are performed at 37°C. In the case of ori-minus SVtsA58 DNA, serial passages of cells are done at 35°C.
4. Add DNA Ca-P co-precipitates to each dish and incubate for 4–16 h.
5. Remove the precipitate and wash cells 2× with PBS. Re-feed cells with the appropriate medium.
6. Allow cells to reach confluence. Change medium twice weekly. Continue to incubate cultures until foci appear. Then calculate the transformation frequency.
7. Resuspend cells once foci are clearly visible. Subculture at high density (1:3–1:5 ratios or $1–3 \times 10^6$ cells/100 mm dish). Culture all cells in order to maintain all transformants.
8. Allow cultures to attain confluence while monitoring the reappearance of foci. Freeze replicate cultures under conditions for viable-cell storage.
9. Subculture cells repeatedly at high cell density until the normal cells in the culture become senescent. The culture becomes 100% large T antigen-positive at this time.
10. Eventually the cultures of the transformed cells exhibit a gradual decrease in the rate of growth, indicating the entry into crisis. At this point, subculture at 1:3 rather than 1:5 ratios, without discarding any cells. In the absence of cell proliferation, maintain cultures without subcultivation.
11. Passage cells which survive crisis at high cell densities to ensure that immortal cells are actively growing.
12. Freeze immortal cells under conditions for viable-cell storage at early passage for reference.
13. Verify that cells are 100% positive for large T antigen protein by immunohistochemical methods. It should be noted the immortal cells may be derived from more than one SV40 transformant in steps 3–7.

In the event that immortal cell lines are not isolated following crisis, an earlier passage of cells (step 8) should be thawed and steps 9–13 repeated with larger cell numbers. In some cases it may be desirable to isolate immortal cells from individual transformants. Multiple foci should be picked at step 7 and passaged because some individual transformants may not yield immortalized cells at detectable frequences.

Once the culture of immortal cells is obtained, the cells can be cloned. The cloned sublines can be assessed to determine whether they are derived from different original SV/HF based on Southern blot analysis of the integrated SV40 genome. Different pre-immortal SV/HF would be expected to have different patterns of SV40 sequences. We have shown that rearrangements of the SV40 genome are not required for immortalization. Hence cell lineage can be verified under conditions where a few integrated copies are involved, such as in origin-minus transformants. In those cases where matched pre-immortal and immortal cell lines are required, multiple independent SV/HF should be tested in parallel; otherwise, the methods are similar.

The molecular basis for immortalization is unknown although some intriguing information is available. Comprehensive analysis is required, and approaches assessing differential gene expression (e.g. cDNA libraries) in immortal versus pre-immortal SV/HF are in progress. One example involves Mn-superoxide dismutase (SOD) which maps to 6q25 and is localized to the mitochondrial matrix. Chromosome rearrangements involving 6q have been reported in human melanoma (and other tumours) and introduction of chromosome 6 has been reported to suppress tumour (melanoma) formation in nude mice. Data on long-term cultivation in cell culture of such hybrids has not been reported. It was recently demonstrated that the expression vectors encoding cDNA for Mn-SOD (but not the anti-sense cDNA) were growth inhibitory when overexpressed in melanoma cells. It has also been reported that immortal SV/HF have reduced levels of Mn-SOD activity but not Cu-SOD (encoded on chromosome 21). The level is reduced to a lesser extent prior to the appearance of the 6q deletion, consistent with a mutation in Mn-SOD. It should be noted that several changes affecting detoxification of oxygen derivatives are associated with such cells, including diminished catalase and glutathione peroxidase. It is possible that the decreased antioxidant capacities are related to alterations of cellular organelles. Decreased Mn-SOD has also been reported in non-SV40 immortalized cells.

5. Conditional SV40 transformants

Temperature-sensitive SV40 constructs have been particularly useful in dissecting processes important for normal cell growth as discussed briefly above. Furthermore, in many cases, cells not only cease to grow when shifted to the non-permissive temperature but also resume expression of normal differentiated properties. This aspect of conditional mutants of large T antigen makes it attractive for the study of differentiation pathways in diverse cell types, as the cell line at 35°C serves as the control for the one shifted to 39°C. The strategies outlined in this chapter for the preparation of SV40 transformed fibroblasts with extended lifespan or immortalized cells (*Protocol 3*) can be exploited with other human cell types which are difficult to study in this regard due to the limited ability to replicate these cells in culture. Human placental cells (infected with tsA mutant virus), thyroid epithelial cells (transfected with tsA DNA) and endometrial cells (transfected with origin-minus tsA DNA) are examples of cell lines which have been developed in this manner. The value of these cell lines is further underscored by the observation that transformation by SV40 may not necessarily lead to an undifferentiated phenotype. tsSV40 transformed cytotrophoblasts display properties of differentiation characteristic of early gestation regardless of the temperature of growth. For example, these cells expressed various cell adhesion molecules, metalloproteinases and hormones indicative of cytotrophoblast differentiation at 35°C. Additional properties such as expression of chorionic gonadotrophin were induced upon temperature shift to 39°C. An even more extensive series of rodent cell lines have been established in culture after introduction of a SV40 *tsA* mutant genome, using virus vectors or DNA transfection. These include rat cerebellar and other central nervous system precursor cells, endometrial cells, hepatocytes, and macrophages.

Two additional classes of SV40-dependent transformants have been developed in the mouse. Cell lines can be isolated from transgenic mice which are generated using tissue specific regulatory elements and the large T antigen coding region. These cell lines have been useful for the understanding of multi-step carcinogenesis for a variety of tissues; for example, lens, liver, skin, pancreas, kidney, and brain. These studies employed a wild-type SV40 genome. Such tumours and cell lines derived from them have proved useful in cases in which it was difficult to obtain the corresponding normal cell line for study. However, the use of temperature-sensitive mutants of large T antigen in transgenic mice is now providing insights not only into normal cell growth and senescence but also into differentiation. In general, tumours do not occur in such transgenic animals. However, relevant cell lines isolated from these animals express a transformed phenotype including becoming immortal at the permissive temperature in culture and reversion to a normal phenotype at the non-permissive temperature, similar to what happens to cells transformed *in vitro*. In many cases, one or more phenotypic markers of the differentiated cell type also become expressed. Cell lines isolated from transgenic mice generated by introduction of SVtsA58 under the MHC (H-2k^{b}) class I promoter behave in such a manner (18). This promoter is active in different tissues and can be induced for higher expression by exposure to interferon γ. A variety of cell lines with different tissue specificities (fibroblast, thymic epithelial cells, crypt cells from colon and small intestines, etc.) have been established in culture from tissues of H-2k^{b}—tsA58 transgenic mice.

Exploitation of the ability to regulate large T antigen expression or function should therefore be useful for the study of mechanisms required for cellular proliferation and senescence in cell types which under normal circumstances prove difficult to study *in vitro*.

6. Other approaches to immortalization of human cells

Viral oncogenes other than SV40 have also been used to immortalize human cells. DNA viruses such as the papillomaviruses and adenoviruses appear to act through mechanisms of viral gene expression which mimic that of SV40. E1A and E1B viral sequences of adenovirus transform and immortalize fibroblasts and epithelial cells. E1A binds to pRb-1 and immortalizes rodent cells; however, E1B (which binds p53) enhances cell survival and is necessary for full transformation. In the case of human cells, both E1A and E1B are required for immortalization. The gene products E7 (binds pRb-1) and E6 (binds p53) of human papillomavirus (HPV) are required to immortalize human fibroblasts and keratinocytes, whereas E6 appears to be sufficient in itself for immortalization of human mammary epithelial cells. Thus, it is clear that inactivation of the tumour suppressor genes *p53* and *pRb-1* is often necessary at least as a first step to alter and extend the pattern of cell growth. Recently DNA constructs encoding HPV genes have been used as an alternative to SV40 T antigen for the generation of immortal human cell lines. *Protocol 3* can be used with the exception that the appropriate sequences are used for DNA transformation in step 3. Constructs using a promoter derived from a retrovirus LTR (19) or other sources are often used instead of the HPV promoter to obtain higher levels of expression of the HPV E6 and E7 intracellularly.

The classes of viruses generally utilized for immortalization of lymphoid cells have been different from those discussed previously for fibroblasts and epithelial cells. Epstein–Barr virus (EBV), a member of the Herpes viruses, has been extensively used to induce immortalization of B lymphocytes (20) for studies on human gene mapping and other purposes. This process is dependent on the expression of the EBV nuclear antigen 2 (*EBNA-2*) gene. *EBNA-2* is a transcriptional regulator which mediates EBV latency gene expression and cellular genes (*cfgr* and *CD23*), most likely accounting for its crucial role in immortalization. In the case of T cells, infection with the human retrovirus T cell leukaemia virus (HTLV) has been used. Immortalization is dependent on expression of the viral tax gene. In each cell system, the isolation of the immortalized lymphoid cell requires methodologies appropriate to the cultivation of these cells in suspension and is therefore different from that described in *Protocol 3*.

It is recommended that all viruses and virus vectors be handled in accordance with safety guidelines recommended by the National Institutes of Health (US) or other agencies.

7. Summary

Immortalization of human cells by SV40 allows for the analysis of processes which affect cellular proliferation. Isolation of genetically matched sets of normal, SV40-preimmortal, and SV40-immortal cells facilitates investigations of mechanisms involved in senescence and immortalization. We have stressed in this chapter the usefulness of SV40 transformed human fibroblasts; however, many different cell types can be similarly immortalized by SV40. There are some instances where other viral agents may be more applicable, as in the case of B lymphocytes which are readily immortalized by EBV.

Although some SV40 transformants do not yield immortal cell lines, such transformants with extended lifespans are none the less valuable for the development of model systems where replication *in vitro* was otherwise exceedingly difficult. Also, in this regard, the utilization of large T antigen function capable of differential regulation often permits the study of cellular differentiation. Thus, transformation by SV40 is applicable for the investigation of the multifacets of normal cell growth and differentiation in human and other mammalian cell systems.

References

1. Hayflick, L. (1965). *Exp. Cell. Res.*, **37**, 614.
2. Radna, R., Caton, Y., Jha, K. K., Kaplan, P., Li, G., Traganos, F., and Ozer, H. L. (1989). *Mol. Cell. Biol.*, **9**, 3093.
3. Wright, W. E., Pereira-Smith, O. M., and Shay, J. W. (1989). *Mol. Cell. Biol.*, **9**, 3088.
4. Hubbard-Smith, K., Patsalis, P., Pardinas, J. R., Jha, K. K., Henderson, A., and Ozer, H. L. (1992). *Mol. Cell. Biol.*, **12**, 2273.
5. Goldstein, S. (1990). *Science*, **249**, 1129.
6. Wright, W. E. and Shay, J. W. (1992). *Trends Genet.*, **8**, 193.
7. Stein, G. H., Beeson, M., and Gordon, L. (1990). *Science*, **249**, 666.
8. Girardi, A. J., Jensen, F. C., and Koprowski, H. (1965). *J. Cell. Comp. Physiol.*, **65**, 69.
9. Neufeld, D., Ripley, S., Henderson, A., and Ozer, H. L. (1987). *Mol. Cell. Biol.*, **7**, 2794.
10. Lin, J.-Y. and Simmons, D. T. (1991). *J. Virol.*, **65**, 6447.
11. Pereira-Smith, O. M. and Smith, J. R. (1988). *Proc. Natl. Acad. Sci. USA*, **85**, 6042.
12. Ray, F. A. and Kraemer, P. M. (1992). *Cancer Genet. Cytogenet.*, **59**, 39.

13. Sandhu, A. K., Hubbard, K., Kaur, G. P., Jha, K. K., Ozer, H. L., and Athwal, R. S. (1994). *Proc. Natl. Acad. Sci. USA*, **91**, 5498.
14. Ning, Y., Weber, J. L., Killary, A. M., Ledbeter, D. H., Smith, J. R., and Pereira-Smith, O. M. (1991). *Proc. Natl. Acad. Sci. USA*, **88**, 5635.
15. Hensler, P. J., Annab, L. A., Barrett, J. C., and Pereira-Smith, O. M. (1994). *Mol. Cell. Biol.*, **14**, 2291.
16. Ogata, T., Ayusawa, D., Namba, M., Takahashi, E., Oshimura, M., and Oishi, M. (1993). *Mol. Cell. Biol.*, **13**, 6036.
17. Bischoff, F., Yim, S. O., Pathak, S., Grant, G., Siciliano, M. J., Giovanella, B. S., Strong, L. C., and Tainsky, M. A. (1990). *Cancer Res.*, **50**, 7979.
18. Jat, P. S., Noble, M. D., Ataliotis, P., Tanaka, Y., Yannoutsos, N., Larson, L., and Kioussis, D. (1991). *Proc. Natl. Acad. Sci. USA*, **88**, 5096.
19. Klingelhutz, A. J., Barber, S. A., Smith, P. A., Dyer, K., and McDougall, J. K. (1994). *Mol. Cell. Biol.*, **14**, 961.
20. Middleton, T., Gahn, T. A., Martin, J. M., and Sugden, B. (1991). *Adv. Cancer Res.*, **40**, 19.

Appendix. Changes identified in senescing mammalian cells

A. Bio markers

	Changes during senescence	Reference
L7 (ribosomal protein) (pHE7)	mRNA decreases	Seshadri *et al.* (1993). *J. Biol. Chem.* **268**, 18476
SAG	Increases	Wistrom and Villeponteau (1992). *Exp. Cell Res.* **199**, 355
Ferritin Heavy Chain	Increases	Thweatt *et al.* (1992). *Exp. Gerontol.* **27**, 433
Plasminogen Activator Inhibitor	Increases	Kumar *et al.* (1992). *Proc. Natl. Acad. Sci. USA* **89**, 4683
Terminin	Increases	Wang and Tomaszewski (1991). *J. Cell Phys.* **147**, 514
Statin	Increases	Wang *et al.* (1989). *Exp. Gerontol.* **24**, 485
Cathepsin B	Decreased enzyme activity	DiPaolo *et al.* (1992). *Exp. Cell Res.* **201**, 500
EF1-alpha	Protein and activity decrease	Cavaullius *et al.* (1986). *Exp. Gerontol* **21**, 149
Calmodulin	Uncoupled from cell cycle	Brooks-Frederich *et al.* (1992). *Exp. Cell Res.* **202**, 386

	Changes during senescence	Reference
EPC-1	Decreases	Pignolo *et al.* (1993). *J. Biol. Chem.* **268**, 8949
DNA polα	Decreases	Pendergrass *et al.* (1991). *Exp. Cell Res.* **192**, 418
PCNA	Decreases	Chang *et al.* (1991). *J. Biol. Chem.* **266**, 8663
LPC-1	Increases	Cristofalo and Pignolo (1993). *Phys. Rev.* **73**, 617
poly (ADP-ribose) polymerase	Activity decreases	Dell'Orco and Anderson (1991). *J. Cell Phys.* **146**, 216
ODCase	Decreases	Chang and Chen (1988). *J. Biol. Chem.* **263**, 11431

B. Extracellular matrix proteins

Hyaluronic acid	Uptake is reduced	Adolph *et al.* (1993). *J. Surg. Res.* **54**, 328
Total collagen	Decreased mRNA	Furth, J. J. (1991). *J. Gerontol.* **46**, B122
Fibronectin	(a) Increases	(a) Kumazaki *et al.* (1991). *Exp. Cell. Res.* **195**, 13
	(b) Alternative splicing	(b) Magnuson *et al.* (1991). *J. Biol. Chem.* **266**, 14654
Metallothionein	Increases	Luce *et al.* (1993). *Exp. Gerontol.* **28**, 17
Collagenase	Increases	Millis *et al.* (1992). *Exp. Cell. Res.* **201**, 373
Stromelysin	Increases	
TIMP-1	Decreases	
F-actin	Increases (neutrophils)	Rao and Cohen (1991). *Mutat. Res.* **256**, 139
Stimulus-induced actin polymerization	Lower (neutrophils)	
Total proteoglycan	Decreases	Eleftheriou *et al.* (1991). *Mutat. Res.* **256**, 127

C. Mitochondria

Deletions	Accumulate	Cortopassi *et al.* (1992). *Proc. Natl. Acad. Sci. USA* **89**, 7370

	Changes during senescence	Reference
MnSOD	Increases	Kumar *et al.* (1993). *Exp. Gerontol.* **28**, 505
Cytochrome *c* oxidase	Increases	
NADH:COQ reductase	Increases	Doggett *et al.* (1992). *Mech. Ageing Dev.* **65**, 239
Cytochrome *b*	Increases	

D. Heat shock family of proteins

Mortalin	(a) Cytosolic form increases (b) Induces cellular senescence	Wadhwa *et al.* (1993). *J. Biol. Chem.* **268**, 22239
Hsp 98, 89, 78, 72, 64, 50, 25	Decreased response	Liu (1989). *J. Biol. Chem.* **264**, 12037

E. Cell cycle

c-*fos*	Decreases	Seshadri and Campisi (1990). *Science* **247**, 205
cdc2	(a) Lack of TNF-α induction (b) Decreased mRNA	(a) Tsuji *et al.* (1993). *Exp. Cell Res.* **209**, 175 (b) Stein *et al.* (1991). *Proc. Natl. Acad. Sci. USA* **88**, 11012
cdk2	(a) Lack of TNF-α induction (b) Decreased protein	(a) Tsuji *et al.* (1993). *Exp. Cell Res.* **209**, 175 (b) Dulic *et al.* (1993). *Proc. Natl. Acad. Sci. USA* **90**, 11034
p42 map-kinase	Not phosphorylated	Afshari *et al.* (1993). *Exp. Cell Res.* **209**, 231
cyclin E-cdk2 complexes	Decreased kinase activity	Dulic *et al.* (1993). *Proc. Natl. Acad. Sci. USA* **90**, 11034
cyclin D1-cdk2 complexes	Unphosphorylated cdk2	
Cyc B	Decreased mRNA	Stein *et al.* (1991). *Proc. Natl. Acad. Sci. USA* **88**, 11012
SDI	Increases	Noda *et al.* (1994). *Exp. Cell Res.* **211**, 90

	Changes during senescence	**Reference**
Histone H2A	Synthesis lower in G_1	Dell'Orco and Worthington (1991). *J. Gerontol.* **46**, B81
Thymidine kinase	Decreased	Chen *et al.* (1989). *Exp. Gerontol.* **24**, 523
CB/tk$^{(CCAAT)}$	Decreased	Pang and Chen (1993). *J. Biol. Chem.* **268**, 2909
Histone 3	mRNA decreases (absent)	Seshadri and Campisi (1990). *Science* **247**, 205
AP-1	Reduced activity	Riabowol *et al.* (1992). *Proc. Natl. Acad. Sci. USA* **89**, 157
pRB	Unphosphorylated form increases	Stein *et al.* (1990). *Science* **249**, 666
c-Ha ras	Increased	Cristofalo and Pignolo (1993). *Phys. Rev.* **73**, 617

F. Growth factors

Glycopeptide growth inhibitor activity	Increases	Macieira-Coelho and Soderberg (1993). *J. Cell Phys.* **154**, 92
EGF/EGF-R	(a) Decreased receptor binding (b) Internalization (c) Receptor protein levels	Reenstra *et al.* (1993). *Exp. Cell Res.* **209**, 118
IGF-1	Decreased mRNA	Ferber *et al.* (1993). *J. Biol. Chem.* **268**, 17883
IGF-Binding protein 3	Increases	Goldstein (1993). *J. Cell. Phys.* **156**, 294
80, 84 and 87 kDa secreted polypeptides	Decreases	Eleftheriou *et al.* (1993). *Biochim. Biophys. Acta* **1180**, 304
Hydrocortisone	Extends lifespan	Kondo *et al.* (1983). *Mech. Ageing Dev.* **21**, 335
Dehydroepiandrosterone	Shortens lifespan	
Estradiol	Shortens lifespan	
Progesterone	Shortens lifespan	
Adenosine	Increased secretion	Ethier *et al.* (1989). *Mech. Ageing Dev.* **50**, 159
Inosine	Increased secretion	
Interleukin-1	Increased	Kumar *et al.* (1992). *Proc. Natl. Acad. Sci. USA* **89**, 4683

G. Chromosomal DNA structure

	Changes during senescence	**Reference**
Chromatin	More thermolabile and reorganized	Macieira-Coelho (1991). *Mutat. Res.* **256**, 81
Extrachromosomal circular DNA	Increases	Kunisada *et al.* (1985). *Mechs. Ageing Dev.* **29**, 89
Telomeres	Shorten	(a) Harley *et al.* (1990). *Nature* **345**, 458 (b) Hastie *et al.* (1990). *Nature* **346**, 866

13

Analysis of the effects of purified growth factors on normal human fibroblasts

PAUL D. PHILLIPS and VINCENT J. CRISTOFALO

1. Introduction

Normal human diploid fibroblast-like cells are widely used for studies of *in vitro* cell ageing and in studies of growth factor action and cell proliferation. Because of their special growth characteristics the manipulation of these cells requires special attention and understanding to ensure the vigorous and optimum growth of the cultures.

Cells from normal human tissues, explanted and grown in tissue culture, have the following characteristics: (i) the cultures possess a finite and predictable proliferative capacity most accurately measured as cumulative number of population doublings (chapter 12); (ii) the replicative life span of a cell line is inversely proportional to the age of the donor; (iii) normal human fibroblast-like cells have never been shown to transform spontaneously in culture, i.e. they have never been shown spontaneously to acquire the indefinite proliferative capacity characteristic of tumour derived or otherwise transformed cells; and (iv) normal human cells can be transformed into cultures with an indefinite proliferative capacity by infection with DNA tumour viruses such as simian virus 40 (SV40) (1, 2, and Chapter 12). There are very few reports of transformation by chemical mutagenesis (3).

For routine purposes, the serial propagation of normal human cells is most readily accomplished using serum-supplemented medium. Although this is generally acceptable for maintaining stock cultures there are associated problems. The use of serum in specific experimental designs can be the source of obvious, as well as not so obvious, complications. Serum is an extraordinarily complex biological fluid composed of literally hundreds of bioactive components whose effects on any particular cellular response are, at best, not fully understood. Because of the vast number of known and unknown factors in serum it is obvious that a defined mitogen supplement would be preferred. There have been reports of the serial propagation of normal human fibroblasts in completely serum-free (i.e. free of undefined biological fluids), growth factor supplemented medium (4). However, many normal human cell lines require complex and difficult to prepare lipid supplements, in addition to various growth factors, to ensure cell viability and proliferation following subcultivation. Also, subcultivation using proteolytic enzymes (Trypsin) requires protease inhibitors for the cells to proliferate. To avoid many of these problems we have developed a system which supports multiple rounds of cell proliferation in a serum-free medium supplemented with various growth factors (5, 6). This chapter describes primarily our procedures for maintaining cultures of normal human fibroblasts throughout their *in vitro* lifespan in serum-supplemented medium and our procedures for taking stock cultures of these cells and subcultivating them into a serum-free medium for studies of cell ageing and growth factor action.

The serum-free serial subcultivation of various lines of normal human fibroblast-like cells has been reported. To date, it is necessary either to supplement the medium with a complex lipid mixture that is delivered in the form of liposomes (4) (which are unstable and must be prepared immediately before use) or the medium must be supplemented with very large quantities of bovine serum albumin (BSA) (7) (this acts as a lipid carrier, but also contains many unknown and known mitogens such as insulin-like growth factor-I). In our laboratory we have focused on the short-term (up to 12 population doublings) growth of WI-38 cells in a defined, serum-free, growth factor supplemented medium (5, 6). This allows us to take advantage of the relative ease of carrying our stock cultures in an FBS supplemented medium while permitting us to switch over to the chemically defined serum-free system at any point in the lifespan of the cultures in order to study ageing and specific growth factor action.

We have developed a classification system based on the functional equivalency of mitogens that regulate WI-38 cell proliferation. Analysis of the proliferative response of WI-38 cells to nine mitogens delineates three classes of mitogens. Class I includes EGF, FGF, PDGF and thrombin; Class II includes IGF-I, IGF-II and insulin; and Class III includes dexamethasone (DEX), hydrocortisone. Any Class I mitogen in combination with any Class II mitogen plus any Class III mitogen stimulates DNA synthesis to the levels observed in 10% serum-supplemented medium (8).

Based on this classification we investigated the action of the different classes of growth factors at different times in G_0/G_1. We found that when EGF, IGF-I and DEX were added to quiescent cells at time 0, the cells began to synthesize DNA by about 12 h. When IGF-I and DEX were added at time 0, and EGF then added at 6, 9, or 12 h, there was a delay in the entry into DNA synthesis that was approximately equal to the time for which EGF was withheld. There is a similar pattern when EGF and DEX are added at time 0, and IGF-I is then added at 6, 9, or 12 h. The pattern is different, however, for DEX. DEX can be added to EGF and IGF-I at 6, 9 or 12 h without delaying the time of entry or affecting the magnitude of the response.

Other work from our laboratory has shown that the EGF, IGF-I, and PDGF receptor systems are largely unchanged with age although some subtle modifications do occur (9, 10 and see below). For instance, the tyrosine kinase activity of EGF receptors isolated from senescent cells is extremely labile when exposed to non-ionic detergents such as NP40, whereas its activity is unchanged from that of young cells when assayed in plasma membrane preparation or *in situ* (11–15).

2. Studies with growth factors

2.1 Materials

2.1.1 Growth factors

WI-38 cells are responsive to a fairly large number of different mitogens, but maximum stimulation (i.e. equivalent to 10% FBS supplemented medium) requires specific combinations of growth factors. In the following sections we will describe our methods for growing WI-38 cells through at least 10 population doublings at a rate and to an extent equivalent to 10% FBS supplemented medium. Our original growth factor formulation consisted of the following:

- partially purified platelet-derived growth factor (PDGF) at 3 μg/ml;
- epidermal growth factor (EGF) at 100 ng/ml;
- insulin (INS) at 5 μg/ml;
- transferrin (TRS) at 5 μg/ml (all from Collaborative Research Inc.);
- dexamethasone (DEX) at 55 ng/ml (Sigma).

Although some of these preparations (PDGF, INS, TRS) were only partially purified, this mixture provides an excellent starting point for growth studies.

2.1.2 Growth medium and trypsinizing medium

We use a modified form of the basal medium MCDB-104 (14) (formula number 82-5006EA GIBCO) which differs from the original formulation as follows: Na pantothenate is substituted for Ca pantothenate, without $CaCl_2$, without Hepes buffer, and without Na_2HPO_4. This formulation allows us to use two different buffer systems and to work with varying concentrations of $CaCl_2$. We have this medium prepared as a powdered mixture in packets, each sufficient to make up 1 litre of liquid medium. To 700 ml of deionized distilled water add one packet of the powdered medium. Add the following additional components in the order listed:

(i) 1.0734 g $Na_2HPO_4 \cdot 12H_2O$
(ii) 1.754 g NaCl

(iii) 1.0 ml of 1 M $CaCl_2$

(iv) 1.176 g $NaHCO_3$

(v) Bring this to 1 litre with deionized-distilled water.

(vi) Filter sterilize the medium through a 0.22 μm pore filter into sterile glass bottles.

(vii) If the medium has a purple tint, it indicates that it has become alkaline (pH above about 7.6). This should be corrected by bubbling a stream of 5% CO_2: 95% air through the medium after filter sterilization.

Do not attempt to freeze this medium because as it freezes the pH rises. Upon thawing the medium, salts will occasionally have precipitated out of solution and may not redissolve. Properly prepared medium should have a pH of 7.3–7.5. We prepare the medium in small batches and keep it refrigerated for up to three weeks. If it is impractical to buy the powdered medium it can be prepared from the original formulation or our modification by following the procedures described in ref. 16.

A variation of MCDB-104 must be made up to be used for the concentrated growth factor stock solutions. To 900 ml of deionized distilled water add one packet of the MCDB-104 and the following:

(i) 1.0743 g Na_2HPO_4 $12H_2O$

(ii) 11.9 g Hepes buffer

(iii) 1.0 ml of $CaCl_2$

(iv) 25.0 ml of 1 M NaOH

(v) Adjust the pH to 7.5 by titration with 1 M NaOH.

(vi) Bring the volume to 1 litre with deionized-distilled water.

(vii) Sterilize by filtration through a 0.22 μm filter.

This Hepes buffered medium can be stored frozen until needed. This is the original formulation of MCDB-104 (16). Although WI-38 cells seem to tolerate the 50 mM Hepes it causes the formation of intracellular vacuoles (17), the significance of which are unknown. Other cell lines may be even more sensitive to it. For these reasons we have adopted a bicarbonate-based medium for cell growth and only use the Hepes-based medium to make up 100× growth factor stock solutions.

Make up the growth factor stock solutions in the Hepes buffered form of the medium, using sterile plastic pipettes, and store the peptide growth factors in sterile plastic test tubes. DEX should be stored in sterile siliconized glass test tubes. EGF at 2.5 μg/ml (100×), INS at 500 μg/ml (100×), TRS at 500 μg/ml (100×), and 5 mg/ml DEX in 95% ethanol then diluted to 5.5 μg/ml with the MCDB-104 (100×). PDGF must be handled differently because it readily adsorbs to glass surfaces. Follow the individual supplier's instructions and use PDGF at approximately 6 ng/ml if it is highly purified and at about 3 μg if it is only a crude, partially purified preparation. In our laboratory both the human and porcine forms of PDGF are equally potent as mitogens. All of these stock solutions should be dispensed into 0.5 ml to 1.0 ml volumes and stored at −20°C for short periods (up to 4 weeks) or at −70°C for longer periods (3–4 months). Finally, prepare a filter sterilized solution of 1 mg/ml soybean trypsin inhibitor Type I-S (1×) (Sigma). We dispense this into 8 ml aliquots into sterile plastic tubes. Avoid repeated freeze–thaw cycles for all stock solutions.

Prepare the exact amount of serum-free, growth factor-supplemented medium that is required for the number of tissue culture vessels that are to be seeded with cells. This medium must be prepared in a sterile plastic container such as a 50 ml capped centrifuge tube, or a tissue culture flask. Using a sterile plastic pipette or sterile plastic graduated cylinder dispense the appropriate volume of medium into the tissue culture vessels at 0.53 ml/ cm^2. Close the caps on the vessels and lay the vessels on the floor of the laminar flow hood with the cell growth surface down.

2.2 Trypsinizing and harvesting the cells

Once the cells are released the procedure changes. Add 8 ml of the soybean trypsin inhibitor solution to the flask (this is instead of 8 ml of complete MEM, FBS-containing medium), and pipette gently up and down in order to form a single cell suspension. Remove a 0.5 ml aliquot of the cell suspension and count the number of cells.

The cells must now be removed from the soybean trypsin inhibitor solution. Wash the cell harvest once by centrifugation at 75 *g* for 5 min at 4°C. Resuspend the cells in 7–10 ml of serum-free growth factor-free $NaHCO_3^-$ buffered MCDB-104. Only 50–70% of the original cell number are recovered after the centrifugation. Higher speeds or longer centrifugation times may recover more cells but there will be a decrease in viability. Therefore, we recommend the short low speed spin described above. Remove a 0.5 ml aliquot from the resuspended cell harvest and count the number of cells. This will allow for the proper seeding density to be achieved. With experience you will be able to estimate the cell loss during centrifugation and then be able to judge the appropriate volume in which to resuspend the cells so that: (i) the cells will be concentrated enough to give reliable counts on the Coulter Counter, and (ii) the inoculum volume will be small compared to the volume of growth factor-supplemented medium in the flasks and thus have only a minimal dilution effect (5% or less) on the concentrations of growth factors. Then, using a plastic pipette seed the flasks as described in Section 2.2. Follow this by gassing the flasks with the 5% CO_2: 95% air mixture, close the caps tightly and place in the CO_2 incubator (Section 2.2).

For studies that involve growth curves it is both convenient and economical to grow the cells in multiwell plates (e.g. 24-well plates). Essentially the same procedure is used in this case. The wells are approximately 2 cm^2 and receive 1.0 ml of growth factor-supplemented medium. When the cells are resuspended after the centrifugation they must be resuspended such that there are 2×10^4 cells per 25 l to 50 l. Then, using a variable micropipette (e.g. Eppendorf or Rannin) which has been wiped down with 70% ethanol, and a sterile tip inoculate the wells with the appropriate number of cells. Place the multiwell plates in the 37°C CO_2 incubator.

2.2.1 Alternate growth factor formulations

In the preceding section we described our original growth factor formulation for culturing WI-38 cells under serum-free conditions. The combination of PDGF, EGF, INS, TRS, and DEX supports growth at a similar rate and to a final cell density as does 10% FBS supplemented medium. A major drawback to this formulation is the expense of even partially purified PDGF. The major effect that PDGF has on these cells, as well as other normal human fibroblasts, is to drive them to a higher saturation density. PDGF can be left out of the medium with very little effect upon the growth rate for the first few days in culture. Without PDGF the final saturation density is about 70% (7×10^4 cells/cm^2) of that obtained with it (1×10^5 cells/cm^2). Since EGF, INS, TRS, and DEX support low density cell growth as well as FBS, there is an added advantage to leaving PDGF out of the formulation. There are fewer mitogenic components and therefore it is a less complex system to study.

INS is an effective mitogen for many cell types, and it is now known that the mitogenic action of INS is derived from its ability to bind, at high concentration (and low affinity), to the insulin-like growth factor-I (IGF-I) receptor (18). The IGF-I receptor actually mediates the mitogenic response to both of these growth factors. IGF-I is maximally effective with WI-38 cells at 100 ng/ml, i.e. at a 50-fold lower concentration than INS. IGF-I can be freely substituted for INS in the growth factor mix, however, it is also very expensive and comparable growth can be obtained using INS [10].

Finally, it is possible to eliminate the TRS from the serum-free medium. To do this a $FeSO_4$ stock solution must also be prepared. Make up a 1 litre solution of 1 mM $FeSO_4$ (200 ×) and add 1 drop of concentrated HCl. Filter sterilize this through a 0.22 μm filter. This solution can be stored at room temperature for approximately 2 months. Discard the solution at the first hint of colour change (16). This solution is used to supplement the basal medium just before the growth factors are added. In our original formulation we used TRS, the plasma iron transport protein, as the vehicle for getting iron into the cells. Subsequently we and others (6, 19) observed that if slightly acidic $FeSO_4$ is added to the medium then TRS is not required. This is apparently dependent on the oxidation state of the iron. In the ferrous state it apparently passes freely into the cell, but when it is oxidized to the ferric state (which occurs spontaneously with time) it requires the transport protein TRS in order to enter the cell.

2.2.2 Special precautions for serum-free cell growth

Although serum-containing medium contributes a large number of unknown variables to the tissue culture system it also provides protection for fastidious cells. For example, serum can adsorb heavy metals or other contaminants which may be found in the deionized water used to prepare the basal medium. Therefore, water quality becomes critical when cells are switched over to a serum-free system such as the one described above. We use deionized, glass-distilled water for our serum-free experiments and find it satisfactory.

It is important that the cells be harvested using the soybean trypsin inhibitor rather than FBS-containing medium (FBS contains trypsin inhibitors). This is because mitogens present in the serum may be sequestered inside the cells or associated with their plasma membranes. The presence of even small quantities of serum complicates the interpretation of cell growth response data.

The other important consideration when working with serum-free, growth factor-supplemented medium is ensuring that the growth factors are not lost by adsorption to glass surfaces. We use only sterile plastic, centrifuge tubes, pipettes, flasks, etc. when handling peptide growth factors. Some, such as EGF, are relatively easy to work with, whereas others such as IGF-I and PDGF are extremely 'sticky' and must be handled only with plastic pipettes and tubes and with the minimum of manipulations.

2.3 A classification of growth factors for WI-38 cells

There are a number of growth factors which stimulate WI-38 cell proliferation, and these can be functionally grouped into three classes (8) as shown in *Table 1*. Class I includes EGF, fibroblast growth factor (FGF), PDGF, and thrombin (THR). Class II includes IGF-I, IGF-II (or the rat homologue multiplication stimulating activity, MSA), and INS. Class III includes hydrocortisone or its synthetic analogue DEX. At low cell density, members of each of the three classes act synergistically in stimulating cell proliferation. Any Class I mitogen in combination with any Class II and either Class III mitogen stimulate growth to an extent similar to FBS-supplemented medium. By using this scheme it should be possible to employ a variety of growth factors in order to maximally stimulate various types of human fibroblast cultures derived from a variety of anatomical sites.

We have been systematically working from the outside in, so to speak, in order to determine how these systems operate and identify age-associated changes which may play a role in the loss of growth factor responsiveness. We have studied, for a series of growth factors, cell responsiveness and receptor binding, and autophosphorylation.

We are interested in the specific mitogens which regulate these and other proliferation associated activities. *Table 1* illustrates a variety of growth factors which stimulate DNA synthesis in young cells. Low density mitogen-

Table 1. The mitogenic functional equivalency of EGF, FGF, PDGF and thrombin

	EGF (25 ng/ml)	FGF (100 ng/ml)	PDGF (6 ng/ml)	THR (500 ng/ml)	IGF-I (100 ng/ml)	DEX (55 ng/ml)	% LN [a]
1	+	+	+	+	+	+	58
2	+	−	−	−	+	+	51
3	−	+	−	−	+	+	60
4	−	−	+	−	+	+	53
5	−	−	−	+	+	+	60
6	+	−	−	−	+	−	19
7	+	−	−	−	−	+	20
8	−	−	−	−	+	+	20
9	+	−	−	−	−	−	18
10	−	−	−	−	+	−	18
11	−	−	−	−	−	+	16
12	−	−	−	−	−	−	9
13	10 % fetal bovine serum						59

[a] Per cent labelled nuclei.
Low density cultures of young cells (<50% life span completed) were made quiescent by re-feeding with MCDB-104. After 48 h they were re-fed as shown along with 1 Ci/ml [^{3}H]TdR. After 24 h they were fixed and prepared for autoradiography and triplicate coverslips were scored for each condition.

deprived young cells were re-fed with various combinations of growth factors together with [^{3}H]TdR. After 24 h they were fixed and prepared for autoradiography, and the percentage of labelled nuclei were scored. Combining EGF, FGF (fibroblast growth factor), PDGF, thrombin, IGF-I (insulin-like growth factor-I) and DEX gives a proliferative response equivalent to serum (compare 1 and 13 in *Table 1*). Equally effective, however, are IGF-I and DEX in combination with EGF (2), FGF (3), PDGF (4), or thrombin (THR) (5). By using EGF as a representative of these last four factors and testing various combinations of EGF, IGF-I, and DEX, however, we see that when any one is left out the stimulation is barely more than basal medium alone (compare 6–8 with 12 in *Table 1*). This points to a synergistic effect. Note that FGF, thrombin and IGF-I are also potent inducers of DNA synthesis. These are in addition to the previously mentioned growth response to EGF, PDGF, INS, and DEX.

These and related experiments have been used to construct a classification scheme. The factors which we studied that stimulate cell proliferation in WI-38 cells can be placed into three classes as shown in *Table 2*. The Class I mitogens include EGF, FGF, PDGF, and THR. These growth factors act via their own separate cell surface receptor systems (20–23). The Class II mitogens IGF-I (also known as somatomedin c) IGF-II (or the rat homologue multiplication stimulating activity) and INS. These structurally related factors, however, all act by their varying abilities to bind to the IGF-I receptor (18). Binding to their own receptors on these cells, however, does not medi-

Table 2. Mitogen classification for WI-38 cells

Class I	**Class II** [b]	**Class III** [c]
EGF [a]	IGF-I	HC [d]
FGF [a]	IGF-II	DEX
PDGF [a]	(MSA)	
Thrombin [a]	Insulin	

[a] Separate Class I receptor systems.
[b] Function mitogenically through the IGF-I receptor system.
[c] Function mitogenically through the glucocorticoid receptor system.
[d] Hydrocortisone.

ate cell proliferation. The Class III mitogens are made up of hydrocortisone or the synthetic analogue DEX. Both of these steroids operate through the glucocorticoid receptor system (24) in WI-38 cells.

The Class I mitogens are functionally equivalent in that when any one of them is combined with a Class II mitogen and a Class III mitogen then DNA synthesis is stimulated to an extent equivalent to fetal bovine serum. This means that for the maximum proliferative response we must activate the glucocorticoid receptor system, the IGF-I receptor system and any one of several other receptor systems.

The optimal concentrations for these growth factors are: EGF at 25 ng/ml, FGF at 100 ng/ml, PDGF at 6.6 ng/ml, partially purified THR at 500–1000 ng/ml, IGF-I at 100 ng/ml, IGF-II at 400 ng/ml INS at 5 g/ml and DEX at 55 ng/ml. In our experiments we routinely use EGF as the Class I mitogen, IGF-I as the Class II mitogen and DEX as the Class III mitogen.

In fact, these three factors support multiple rounds of cell proliferation. *Table 3* shows that EGF, IGF-I, and DEX are as effective as EGF, INS, and

Table 3. Young cell growth supported by insulin or IGF-I

		Cells/cm^2 on day 7
EGF insulin	DEX	7.4×10^4
EGF IGF-I	DEX	7.6×10^4
EGF	DEX	2.0×10^4

Cultures of cells (<50% LSC) were seeded at 1×10^4 cells/cm^2 into each of the growth factor combinations in MCDB-104. After 7 days the cell number on triplicate wells for each set were determined by means of a Coulter counter.

DEX in supporting growth. This is to a significantly greater extent then EGF and DEX alone.

We have examined all three growth factor receptor systems in some detail. For binding there is no age-associated change in the EGF (9) or IGF-I (25) receptors, although there is some decrease in the amount of DEX binding (24). However, whether this is sufficient to account for the decreased responsiveness is still unknown. There are no changes in any of the ligand-receptor affinities with age. The EGF and IGF-I receptors are known to be tyrosine specific autocatalytic protein kinases (26, 27). For EGF receptors in membrane preparations this enzyme activity appears to be unchanged with age (13, 28). However, when the EGF receptor is purified by detergent solubilization, the autocatalytic kinase activity in senescent cell preparations is greatly reduced and nearly absent (15). At present we are pursuing the cause of this age-associated increased enzyme lability but at this time neither its basis or functional significance is understood. We have only recently begun studies of the tyrosine kinase activity of IGF-I receptor and have not drawn any conclusions at this point. And finally, this enzyme activity is not applicable to the DEX receptor.

Given what we know about these different receptor systems we can say that, at least as far as the EGF and IGF-I receptors are concerned, changes in receptor number or binding affinity cannot account for the loss of proliferative responsiveness. As far as their tyrosine kinase activities are concerned this is still unclear. But at least for the EGF receptor it is not a simple case of a loss of enzyme activity under all experimental conditions. A decrease in the number of DEX receptors may or may not be significant.

Recently we have begun a series of studies designed to characterize the PDGF receptor system in cultures of young and senescent cells. We find that old cells can bind at least twice as much ^{125}I-labelled PDGF as young cells. This is consistent with the increase in cell size with age, and similar to what we have observed for EGF and IGF-I specific binding. The K_d of the PDGF-receptor complex is approximately 2×10^{-9} M, and does not change with age. This is also similar to what we have observed for the EGF and IGF-I receptor systems. The PDGF binding data are summarized in *Table 4*.

We have also recently investigated whether the PDGF receptor becomes phosphorylated on tyrosine, in response to PDGF, in membranes prepared from young and old cells. Under these conditions, the PDGF-stimulated phosphorylation of the receptor appears to be equivalent in membranes from young and old cells. This is the same as has been observed for the EGF receptor (13).

The properties of the EGF, IGF-I, PDGF, and DEX receptor systems which we have studied are summarized in *Table 5*.

As we described earlier there is a variety of G_0/G_1 events which occur in response to mitogen stimulation in both young and old cells. These include late G_1 events like increased thymidine kinase activity, thymidine triphosphate

Table 4. ^{125}I-labelled PDGF specific binding to young and old cells

	Young cells	Old cells
Radioactivity speciically bound/10^5 cells (c.p.m.)	450	900
50% displacement of ^{125}I-labelled PDGF	2.8 nM	2.4 nM

Confluent cultures of young (<50% lifespan completed) and old (>90% lifespan completed) cells were incubated at 4 °C with 1 ng/ml of ^{125}I-labelled PDGF and increasing concentrations of unlabelled PDGF. Non-specific binding was determined in the presence of a 100-fold excess of unlabelled PDGF.

Table 5. Summary of age-associated changes in the EGF, IGF-I, PDGF, and DEX receptor systems

	Receptor systems			
	EGF	**IGF-I**	**PDGF**	**DEX**
Binding sites per cell	I	I	I	D
Apparent K_D	NC	NC	NC	NC
Receptor phosphorylation in membranes	NC	ND	NC	NA
Receptor phosphorylation in solution	D	ND	ND	NA

Abbreviations: D decreases with age; I, increases with age; NA, not applicable; NC, no change with age; ND, not done.

pool expansion and histone H3 gene expression. So at least some of these pathways appear complete. There does not appear to be any gross growth factor receptor dysfunction, although certainly some subtle changes do take place.

Given such similarities it is potentially important to know if any sequential, temporal actions exist with respect to the three growth factors. If growth factors act at different times, it may be possible to dissect age-associated changes in responsiveness to specific growth factors. This should make it possible to isolate the pathways involved in the loss of proliferative capacity. With this in mind we have examined the timed addition of each of the three growth factors and monitored the entry of cells into DNA synthesis. *Table 6* shows the results of such an experiment. We took quiescent mitogen-deprived, middle-aged cells and stimulated them with EGF, IGF-I and DEX or various combinations of these factors together with [^{3}H]TdR. At the times

Table 6. The effect of the timed addition of EGF, IGF-I, and DEX on the entry of cells into DNA synthesis

Time of Growth Factor Addition (h)				
0	**6**	**9**	**12**	**Entry into S phase (h)**
EGF IGF-I DEX				9–12
IGF-I DEX	EGF			15–18
IGF-I DEX		EGF		21–24
IGF-I DEX			EGF	21–24
EGF DEX	IGF-I			12–15
EGF DEX		IGF-I		18–21
EGF DEX			IGF-I	21–24
EGF IGF-I	DEX			9–12
EGF IGF-I		DEX		9–12
EGF-IGF-I			DEX	12–15

Low density cultures of young cells were made quiescent by refeeding with MCDB-104. After 48 h they were re-fed as shown along with 1 Ci/ml [^{3}H]TdR. At 3 h intervals duplicate coverslips for each experimental conditon were fixed, prepared for autoradiography and the percentage labelled nuclei were scored. At the times shown EFD, IGF-I or DEX were added.

indicated the cells, which were growing on coverslips, were fixed and prepared for autoradiography and scored for percentage labelled nuclei. When EGF, IGF-I and DEX were added at time 0 the cells began to enter DNA synthesis by about 12 h. When IGF-I and DEX were added at time 0 and EGF then added at 6, 9, or 12 h there was a delay in the entry into DNA synthesis which is approximately equal to the time for which EGF was withheld. There is a similar pattern when EGF and DEX are added at time 0 and IGF-I then added at 6, 9, or 12 h. Although there is a suggestion that some cells enter S phase on time, there is certainly a clear delay for the majority of entering cells which is approximately equal to the time for which IGF-I was withheld. The pattern changes, however, for DEX. DEX can be added to EGF and IGF-I at 6, 9, or 12 h without delaying the time of entry or effecting the magnitude of the response. It appears that DEX acts as a 'trigger or gate' to send cells, which have otherwise progressed through G_1, on into DNA synthesis. Unlike either EGF or IGF-I, DEX does not have to be continuously present, but only near the G_1/S boundary.

References

1. Cristofalo, V. J. and Pignolo, R. J. (1993). *Physiol. Rev.*, **73,** 617.
2. Jensen, F., Koprowski, H., and Ponten, J. A. (1968). *Proc. Natl. Acad. Sci. USA*, **50,** 343.
3. Zimmerman, R. J. and Little, J. B. (1983). *Cancer Res.*, **43,** 2183.

4. Bettger, W. J., Boyce, S. T., Walthall, B. J., and Ham, R. G. (1981). *Proc. Natl. Acad. Sci. USA*, **78,** 5588.
5. Phillips, P. D. and Cristofalo, V. J. (1980). *J. Tissue Culture Methods*, **6,** 123.
6. Phillips, P. D. and Cristofalo, V. J. (1981). *Exp. Cell Res.*, **134,** 297.
7. Yamane, I., Kan, M., Hoshi, and Minamoto, Y. (1981). *Exp. Cell Res.*, **134,** 470.
8. Phillips, P. D. and Cristofalo, V. J. (1988). *Exp. Cell Res.*, **137,** 396.
9. Carlin *et al.* (1983).
10. Brooks-Frederich *et al.* (1987).
11. Phillips, P. D., Kuhnle, E., and Cristofalo, V. J. (1983). *J. Cell. Physiol.*, **114,** 311.
12. Phillips, P. D., Pignolo, R. J., and Cristofalo, V. J. (1988). *J. Cell. Physiol.*, **133,** 135.
13. Brooks, K. B., Phillips, P. D., Carlin, C. C., Knowles, B. B., and Cristofalo, V. J. (1987). *J. Cell. Physiol.*, **133,** 523.
14. Gerhard, G. S., Phillips, P. D., and Cristofalo, V. J. (1991). *Exp. Cell Res.*, **193,** 87.
15. Carlin, C. R., Phillips, P. D., Knowles, B. B., and Cristofalo, V. J. (1981). *Nature*, **306,** 617.
16. McKeehan, W. L., McKeehan, K. A., Hammond, S. L., and Ham, R. G. (1977). *In Vitro*, **13,** 470.
17. Verdery, R. B., Nist, C., Fujimoto, W. Y., Wight, T. N., and Glomset, J. A. (1981). *In Vitro*, **17**, 956.
18. Van Wyk, J. J., Graves, D. C., Casella, S. J., and Jacobs, S. (1985). *J. Clin. Endocrinol. Metab.*, **61**, 639.
19. Walthall, B. J. and Ham, R. G. (1981). *Exp. Cell Res.*, **134**, 303.
20. Carpenter, G. and Cohen, S. (1976). *J. Cell. Biol.*, **71**, 159.
21. Schreiber, A. B., Kenney, J., Kowalski, W. J., Friesel, R., Mehlman, T., and Maciag, T. (1985). *Proc. Natl. Acad. Sci. USA*, **82**, 6138.
22. Heldin, C.-H., Westermark, B., and Wasteson, A. (1981). *Proc. Natl. Acad. Sci. USA*, **78**, 3664.
23. Carney, D. H., Steirnberg, J., and Fentor, J. W. (1984). *Biochemistry*, **26**, 181.
24. Rosner, B. A. and Cristofalo, V. J. (1981). *Endocrinology*, **108**, 1965.
25. Phillips, P. D., Pignolo, R. J., and Cristofalo, V. J. (1987). *J. Cell. Physiol.*, **133**, 135.
26. Jacobs, S., Kull, F. C., Earp, H. S., Suoboda, M. E., Van Wyk, J. J., and Cuatrecasas, (1983). *J. Biol. Chem.*, **258**, 9581.
27. Carpenter, G., King, L., and Cohen, S. (1978). *Nature*, **276**, 409.
28. Chau, C. C., Geiman, D. E., and Ladda, R. L. (1986). *Mech. Aging. Dev.*, **34**, 35.